Electromagnetic Information Leakage and Countermeasure Technique

Taikang Liu · Yongmei Li

Electromagnetic Information Leakage and Countermeasure Technique

Translated by Liu Jinming, Liu Ying, Zhang Zidong, Liu Tao

国防工業出版社
National Defense Industry Press

Springer

Taikang Liu
The Thirty-third Research Institute of China Electronic Science and Technology Group Corporation
Taiyuan, China

Yongmei Li
Jimei University
Xiamen, China

ISBN 978-981-10-4351-2 ISBN 978-981-10-4352-9 (eBook)
https://doi.org/10.1007/978-981-10-4352-9

Jointly published with National Defense Industry Press, Beijing, China
The print edition is not for sale in China Mainland. Customers from China Mainland please order the print book from: National Defense Industry Press.

Library of Congress Control Number: 2019935159

This Springer imprint is published by the registered company Springer Nature Singapore Pte Ltd.
The registered company address is: 152 Beach Road, #21-01/04 Gateway East, Singapore 189721, Singapore

Foreword

Computers and other information processing equipment are common and pervasive in our daily life, which could cause unintentional radiation, resulting in electromagnetic information leakage and serious threat to information security. In terms of electromagnetic information leakage, analysis of its leakage mechanism and research on its protection countermeasures become a vital issue for ensuring information security. Currently, in domestic, as far as anti-electromagnetic information leakage is concerned, there still lacks work having a complete theoretical system and systematic protection strategy. This book of Electromagnetic Information Leakage and Countermeasure Technique, which is the result of many years of research by authors' team, systematically discusses theory, methods, and new technique on electromagnetic information leakage and its countermeasure.

By analysis of electromagnetic information essence, this book creatively incorporates electromagnetic radiation and propagation theory with information theory and describes the model of electromagnetic information leakage which conforms to the characteristics of electromagnetic information, initially formed a theoretic system on electromagnetic information leakage and countermeasure technique.

Based on the above described theoretic system, starting from the information source, channel, and sink, this book systematically analyzed the mechanism and complete process of generation, spread, interception, and reproduction of electromagnetic information leakage. For each stage, systematic layered protection strategy of anti-leakage, anti-interception, and anti-reconstruction is introduced, as contribute to Electromagnetic Information Leakage and Countermeasure Technique. In addition, many cases and examples of electromagnetic information leakage protection are listed in order to confirm the strategy's scientificity and effectiveness. In the meantime, traditional and new protection material and their application are introduced.

This book exhibits novelty of academic thought with specific and practical content, providing active promote with the development of defense science and technology and information equipment. For electromagnetic information leakage and countermeasure technique and computer technologies, it is an excellent book in terms of multidisciplinary combination. It meets the urgent needs for electromagnetic information leakage protection researchers and engineers; it is also suited for relating researchers and engineers used as a reference book.

Shijiazhuang, China

Liu Shanghe
Academician of the Chinese Academy of Engineering

Preface

In the information age, information has become a critical element influencing a country's development and social progress. Because of further development of information technology in areas like e-commerce, e-government, and military equipment, the problem of information security becomes more and more prominent. For information security, an emphasis is often put into issues including information security of network, U disk data reading, and computer data storage. However, the problem of electromagnetic information leakage by electromagnetic radiation of information equipment is ignored. In fact, for intelligence agency in foreign countries, an important channel of getting information is through electromagnetic information leakage and obtaining confidential information. As more and more information equipment is widely used in the government intelligence agency, military command system and financial field, as well as improvement of the degree of intellectualization of crime, electromagnetic information leakage threat becomes more and more protruding. Therefore, it is very important to carry out research to prevent leakage of electromagnetic information, including establishing electromagnetic information model, and researching on specific protection technique.

By long-term tracking development of international frontier technology and prophase basic research results of electromagnetic information leakage and countermeasure technique, combined with research and accumulation of domestic scientific research institutions, and comprehensive analysis of the latest research progress published at home and abroad, the book systematically describes electromagnetic information leakage and its protection theory, measures, and new techniques.

In recent years, a lot of research work has been carried out by the authors' team aiming at electromagnetic information leakage. Through actual measurement for electromagnetic leakage of main computer components' spectrum and characteristics, including CPU, CRT, LCD/LED, wireless keyboard, wireless mouse, scanner, etc., a large amount of measurement data is collected. Based on electromagnetic leakage characteristic analysis of computer host and its peripheral, the authors' theoretical understanding of electromagnetic information leakage and countermeasure technique got deepened. It also lays the foundation for the research work on electromagnetic information leakage and protection strategy, critical element of interception and reproduction, testing of electromagnetic information leakage, and TEMPEST standards.

Based on electromagnetic radiation and propagation theory and information theory, aiming at electromagnetic information leakage and countermeasure technique, through analysis of large amount of data and experiment, the book forms electromagnetic information leakage model which is more consistent with electromagnetic information characteristics and further introduces reproduction model of electromagnetic information leakage; thus, theoretical system of electromagnetic information leakage and countermeasure technique is initially formed. In perspective of data reconstruction and information reproduction, three core elements are introduced. In particular, timing pulse is analyzed and researched in terms of importance in electromagnetic information security, pointing out that it is interwoven with electromagnetic information leakage, interception, and reproduction technique, which, in data interpreting, is a critical element in data reconstruction and information reproduction. In the book, a systematic layered protection strategy is introduced according to three stages of leakage, interception, and reproduction; in addition, new protection measure and material are presented. By studying the interception and reproduction methods of information leakage caused by electromagnetic emanation, it is found that making the change of electromagnetic radiation as small as possible or keeping the electromagnetic radiation changes constant is a pivotal way to increase the difficulty of interception.

In a word, three innovative points are introduced in this work, the first is timing pulse that plays an important role in electromagnetic information security, the second is systematic layered protection strategy on electromagnetic information, and the third is the establishment of the electromagnetic information model. They all reflect the newest research and development trends in electromagnetic information leakage and countermeasure technique.

The book is divided into nine chapters:

Chapter 1 summarizes the main contents of the whole book, introduces research background in home and abroad, and discusses the threat of electromagnetic information leakage. And it also briefly introduces electromagnetic information interception and reproduction methods, the importance of anti-electromagnetic information leakage and protective measures, and testing standards.

Chapter 2 describes the basis of electromagnetic radiation and propagation theory, electromagnetic wave generation and radiation leakage mechanism. The electromagnetic information leakage model is established based on information theory and electromagnetic theory, and the electromagnetic information leakage limiting factor is defined, and its influencing factors and value ranges are analyzed. Under noise condition, reconstruction and reproduction model is suggested; the quality factor of which is introduced. In the end, electromagnetic information is classified according to effectiveness and mode of communication.

Chapter 3 discusses the objective of interception and processing technique of intercepted data. It is revealed that three core elements of data reconstruction and information reproduction are volume of intercepted information, timing pulse, and data transmission format. The relating cases are analyzed.

Chapter 4 systematically presents the layered protection strategy of electromagnetic information leakage, which includes "anti-leakage", "anti-interception", and "anti-reconstruction". Based on traditional protection technique, a newer technique is described with case form.

Chapter 5 introduces the protection material of electromagnetic information leakage. Besides traditional material, new protection material such as photonic crystal, left-handed material, and carbon fiber composites are introduced with application cases.

Chapter 6, based on analysis of large amount of measurement data, describes the characteristics of electromagnetic information leakage among core component in computer system and gives corresponding protection strategy and measures. Regarding different components in computer system, electromagnetic information leakage and protection are researched accordingly.

Chapter 7 starts from features of electromagnetic information signal and data, describes methods of signal processing and data processing, and furthermore, reports methods that are used in the above chapters' cases, which are classified and compared.

Chapter 8 discusses the objective and significance of electromagnetic information leakage testing; introduces the testing environment, equipment, and methods; and discusses safety assessment method for information equipment.

Chapter 9 presents the standards of electromagnetic information leakage and countermeasure technique; discusses standards including EMC, TEMPEST, and their relation; introduces zoning standard for electromagnetic safety and development of TEMPEST standards in the home and abroad; and discusses development trend of TEMPEST standard.

As electromagnetic information leakage and countermeasure technique is a new interdisciplinary subject, considering readers' different knowledge background, this book states the related knowledge from electromagnetic information point of view in order to help the reader get a better understanding of the book's contents. As a technical monograph, the target readers are field researchers and engineers. It also suited for graduate students of related specialties.

The authors' team has many years of experience on conducting research work on electromagnetic protection technology and has been guided and helped by Prof. Liu Shanghe, member of Chinese Academy of Science. In the writing and publishing process of this book, it got the care and support of Prof. Liu as well. Vice Minister Liu Chenghai of Electronic Information Basic Department of the General Equipment Department also gives a great support for our research work, especially in research on the protection of electromagnetic information security and this book's writing. Here we express our sincere thanks altogether.

In the course of the publication of this book, Gou Jingjing, Liu Peiguo, Wang Xietian gave their support and help; Jiang Weiya, Jiang Yun, Tian Yulin, Gao Pengqi, Zhang Dewei, etc., as lab experts or engineers and technicians, helped us in data testing and provided technical support; graduate students including Shi Sen, Zhao Zhiqiang, Li Youchen, Liu Shuquan, etc., participated in early stage's research of related technology, thanks to all of them.

Because of the limited research expertise of authors' team, mistakes in the book are unavoidable, so criticism and corrections are welcomed from readers.

Taiyuan, China Taikang Liu

Contents

Abbreviations

AM	Amplitude Modulation
BBN	Broadband Noise
BER	Bit Error Rate
BW	Bandwidth
CE	Conducted Emission
CM	Common Mode
dB	Decibel
dB (μV)	Decibels Referenced To 1 Microvolt
dB (μV/m)	Decibels Referenced To 1 Microvolt Per Meter
dBm	Decibels Reference To 1 Mill watt
DM	Differential Mode
DS	Damped Sinusoidal
DSB	Double Sideband
EFS	Electric Field Strength
EIRP	Equivalent Isotropic Radiated Power
EM	Electromagnetic
EMC	Electromagnetic Compatibility
EME	Electromagnetic Environment
EMI	Electromagnetic Interference
EMP	Electromagnetic Pulse
EMRADHAZ	Electromagnetic Radiation Hazard
EMV	Electromagnetic Vulnerability
EP	Electronic Protection
ERP	Effective Radiated Power
ESC	Equipment Spectrum Certification
ESD	Electrostatic Discharge
EUT	Equipment Under Test
FCC	Federal Communications Commission
FIM	Field-Intensity Meter
FM	Frequency Modulation

FSAF	Free-Space Antenna Factor
GTEM	Gigahertz Transverse Electromagnetic Cell
HEMP	High-Altitude Electromagnetic Pulse
HIRF	High Intensity Radiated Field(S)
IBW	Impulse Bandwidth
IEC	International Electro technical Commission
IMI	Intermodulation Interference
ISB	Independent Sideband
ISO	International Organization for Standardization
ITE	Information Technology Equipment
le	Antenna Effective Length For Electric-Field Antennas
lem	Antenna Effective Length For Magnetic-Field Antennas
LEMP	Lightning Electromagnetic Pulse
LISN	Line Impedance Stabilization Network
M&TE	Measuring and Test Equipment
MAE	Maximum Allowable Environment
MDS	Minimum Discernible Signal
NATO	North Atlantic Treaty Organization
NEMP	Nuclear Electromagnetic Pulse
NIST	National Institute of Standards and Technology
NSA	Normalized Site Attenuation
OATS	Open Area Test Site
PSD	Power Spectral Density
RADHAZ	Radiation Hazard
RAM	Radar Absorbing Material
RBW	Resolution Bandwidth
RCS	Radar Cross Section
RE	Radiated Emission
RF	Radio Frequency
RFI	Radio-Frequency Interference
RI	Radiated Immunity
RS	Radiated Susceptibility
Rx	Receiver
SSB	Single Sideband
TEM	Transverse Electromagnetic
Zt	Transfer Impedance

Abstract

This book systematically discusses theory, method, and latest technology of EM information leakage and countermeasure. In particular, an EM information leakage model is created by incorporating information theory and EM radiation theory, which conforms to EM information characteristics. In the model, EM information leakage constraint factor, the restricted factors, and the value range are defined and analyzed. Second, this book proposes a reconstruction and reproduction model under noisy environment. In addition, a quality factor is provided to evaluate the quality of reconstruction and reproduction. Finally, the theoretical system for EM information leakage and countermeasure technology is initially formed, and a systematically hierarchical countermeasure strategy is developed based on the proposed model, namely, anti-leakage layer, anti-interception layer, and anti-reconstruction layer.

The book is a suitable reference book for relevant researchers, engineers, and postgraduate students.

Chapter 1
Introduction

In the information age, information has become a key factor affecting a nation's development and social progress. Due to the rapid development of large-scale integrated circuit, computer technology and information equipment, the cognition and definition of information are updating and developing constantly, forms and types of information, its representation and standardization, and methods of information dissemination are continuously enriching. From the perspective of information security, electromagnetic information can be divided into two categories: one is the information with the purpose of the transmission, such as the plain code or the encrypted code transmitted by telegraph equipment; the other one is the natural information of the equipment, the electromagnetic radiation or leakage to space. The former is the electromagnetic information transmitted for communication, which is active and intentional; the latter is the electromagnetic radiation or leakage of equipment, which is passive and unintentional.

With the deeper development of information technology in fields like e-commerce, e-government, and military equipment, information security is becoming more and more important. During information processing, computer and other digital information equipment will inevitably emit electromagnetic waves to the surrounding environment, so the electromagnetic radiation occurs. These electromagnetic radiation signals are complex in spectrum components and contain a lot of useful information, thus raise a serious threat to information security, resulting in leakage of information. By wireless interception, data reconstruction and information replication, sensitive and confidential information would leak out. Moreover, as information intercepting and reproducing process is hidden and no traces are left, information leakage detection becomes very difficult, causing serious leakage of sensitive and confidential information.

Confidential information stealing through electromagnetic leakage of computer and other information equipment has become an important channel to obtain information for intelligence agencies at home and abroad. Despite the strict limitation to the electromagnetic radiation intensity of information equipment regulated by TEMPEST standards [1] (there are multiple interpretations by foreign countries: Telecommunication Electronic Material Protected from Emanation and Spurious

T. Liu and Y. Li, *Electromagnetic Information Leakage and Countermeasure Technique*,
https://doi.org/10.1007/978-981-10-4352-9_1

Transmission Electromagnetic Pulse Emanation or Transient Standard) of different countries, with continuing improvement of special equipment for electromagnetic information interception's sensitivity, resolution, and other related technology, it is impossible to avoid electromagnetic radiation accompanied by electromagnetic information leakage of computer and information equipment. Electromagnetic information security will be more and more serious, greater harms, and even devastating blows will be brought, in particular, to departments in defense and military, governments and banks. The important characteristics of modern warfare are information warfare and electronic warfare, thus electromagnetic information leakage is a crucial factor determining success or failure of the war.

Therefore, the leakage issue of electromagnetic information has attracted more and more attention of experts from relevant industries, who are actively seeking solutions for electromagnetic information protection. The book expounds the characteristics, ways, and modes of electromagnetic information leakage and the corresponding electromagnetic protection measures; tests the real data with mathematical models through analysis and research of a large amount of actual measurement data; establishes mathematical models for electromagnetic information leakage under ideal conditions and conditions with noise, respectively. In order to study the interception and reproduction of electromagnetic information under noisy conditions, five variables for the quality of data reconstruction and information reproduction are given, and the relation between them is established. At the same time, in the discussion of means of interception and methods of reproduction, the key role of timing pulse in information reproduction is emphatically analyzed. The principle of classified protection against electromagnetic information leakage is put forward. In addition, new methods and materials for protection are introduced. On the basis of the research and analysis of anti-electromagnetic information leakage, this book introduces and analyzes the latest overseas research achievements in the field of electromagnetic information protection. In the discussion of the security of the computer system and its core components, comparison and analysis of the effectiveness of existing electromagnetic protection measures are made through actual measurement of a large amount of various parts and components and the introduction of foreign-related cases. In the end, the book discusses the latest TEMPEST standards in details.

1.1 Electromagnetic Information Leakage

In the process of digital information processing, information equipment will produce electromagnetic radiation with information, which will cause leakage of electromagnetic information. Electromagnetic information leakage can be transmitted not only through space, but also along the power line, signal line, ground wire, metal pipe, and so on. Reproducing the original information through analyzing and processing the electromagnetic waves detected and intercepted, will bring very serious latent threat to information security.

1.1.1 Electromagnetic Information Leakage Threat

In working conditions, any electronic equipment, electrical equipment, computer, and its information devices will produce electromagnetic radiation with information leakage in different degree, which is considered to be an unavoidable electromagnetic phenomenon. Among them, all electronic or information devices that can process data and control information have the problem of electromagnetic information leakage. Today, computers and information devices have been widely used in all walks of life. Their electromagnetic radiation and information leakage are pervasive. Therefore the information security problem is increasingly serious, which is caused by intercepting and reproducing information in various political, military, and economic activities by using highly sensitive receiving equipment.

In the early 1960s, international military organizations have realized that the electromagnetic radiation generated by computers could not only interfere with radio reception, but also lead to information leakage. These leaked electromagnetic radiation "carries" sensitive information processed and transmitted by the information equipment, once the spectrum of the electromagnetic radiation is intercepted, information security will be threatened. Nowadays, information processed by computers can even be received and reproduced by some common electronic devices, which further proves the serious and common problem of information leakage by electromagnetic radiation.

Following are a number of typical events concerning electromagnetic information leakage publicly reported abroad, which illustrate serious threat caused for information security. In particular, through a typical example, we found that the information could be easily got even using nonprofessional receiving equipment in the early stage of electromagnetic information leakage.

In the 1960s, when Britain was negotiating to join the European Economic Community (now the European Union), British Prime Minister worried that French President De Gaulle would obstruct in the process, so he instructed the intelligence agency to try to make clear the French's position in negotiations. The intelligence service, having failed in deciphering the diplomatic documents of France, however, succeeded in reproducing the contents of those diplomatic documents by intercepting the electromagnetic radiation signal of the cryptography machine.

Another example is an attempt by the Dutch people, Van Eck in 1986, who made the problem of electromagnetic information leakage to get widespread concerns [2]. In a van by the street, he used a slightly modified ordinary TV, successfully received and reproduced the content displayed on the CRT display of a computer several meters away. This event caused great shock and let people know not only the existence of electromagnetic information leakage in computer, but also to realize that the electromagnetic radiation information could be intercepted and reproduced by common equipment. In fact, in addition to electromagnetic information leakage of the computer display, CPU, motherboard, keyboard, printer, power line, and data line and control line and other transmission lines would cause leakage as well, the difference only exists in the degree of difficulty in intercepting and reproducing of the leaked electromagnetic information.

In 1989, Smulders published an experimental report on intercepting information from computer RS-232 bus [3]. The report shows that he succeeded in intercepting the data passed from nearby RS-232 buses by using an ordinary short wave/FM radio. In this same report, Smulders also claimed that the receiving distance could be greatly increased by more advanced receivers. Other reporters described in detail the feasibility of receiving electromagnetic information of graphics cards, CPU and RAM within a 10-m sphere of a computer.

In a word, the very thorough research has been carried out on relative technology of electromagnetic information leakage in main western developed countries, and the technology of interception and reproduction has reached a fairly high standard. It is reported that the United States may, in some cases, receive and reproduce the content of the civilian computer 1 km away. Owing to governments' strict secrecy, only until the mid-80s, the problem of electromagnetic information leakage was gradually recognized by the general public. Therefore, the leakage of electromagnetic information is extremely harmful to information security. It is of great significance to study the mechanism and ways of the leakage and corresponding protection strategies.

1.1.2 Ways of Electromagnetic Information Leakage

The electromagnetic information leakage refers to the radiation and leakage of the stray electromagnetic energy from the information equipment to space through the equipment cabinet aperture or the guideline. Information leaked from computers and other information devices are usually in two ways: one is to radiate directly in forms of electromagnetic wave, which is called radiation emission. Radiation emission is direct electromagnetic radiation of the stray electromagnetic energy in forms of electromagnetic waves, which is outward produced by components and parts of information devices and passing through the cabinet aperture and various guidelines; the other way is called conduction emission which is conducted through metal pipelines and cables. Conduction emission is the conduction and radiation of the electromagnetic wave containing information through the power line, signal wire, ground wire, and other transmission medium. Generally, the conduction emission process would be accompanied by radiation emission leakage. For example, the data transmission line, power line, and ground wire of the computer system or information devices, could all become the medium of information transmission, in which the media play the role of an antenna to radiate the conducted signals.

1.1.3 Electromagnetic Information Model

The essence of the electromagnetic information model is the description of digital information. Electromagnetic wave is just the carrier of the information transmission. Therefore, the electromagnetic information model discussing the characteristics of electromagnetic information can be established based on the information theory and theory of antenna and electromagnetic propagation.

The electromagnetic information model is classified as the electromagnetic information leakage model under ideal conditions and in noise conditions and interception and reproduction of electromagnetic information under noise conditions, given the limiting factor of electromagnetic leakage and the principle for value selection, the former reflects the working mechanism of the equivalent antenna, the circuit characteristics and receiving antenna sensitivity and many other factors, which shows the quantity relation between the original and the leaked information. The quality of data reconstruction and information reproduction can be measured by the ratio of the intercepted information amount to the total information amount.

1.1.4 Electromagnetic Information Leakage Feature and Measurement

The computer is a typical modern information device. The analysis of the characteristics of electromagnetic information leakage of the core components of the computer system plays an important role in the leakage and protection of electromagnetic information. Owing to the differences in the working mechanism, data format, and communication protocol, there are different characteristics of electromagnetic leakage, which means different testing methods and means should be applied.

The test is to obtain the electromagnetic radiation data of the device tested through certain methods with the help of special equipment. The security level of the information device can be evaluated through testing. In order to ensure the accuracy of the test results, proper test method and appropriate test environment should be chosen. The whole testing process should comply with relevant TEMPEST standards.

The test equipment and environment will affect the reliability and accuracy of the test results. Usually, the test environment consists of two systems: test system and target system. Figure 1.1 shows a test case for electromagnetic radiation from a notebook. The radiation of the target system, marked by a circle, is received by an antenna of the test system. There is no cable connection between the test system and the target system except electromagnetic wave.

Different test methods should be chosen for different test targets accordingly, there are mainly two categories, the time-domain method and the frequency domain method: the time-domain measurement method can be used to obtain the time-domain wave pattern, the frequency-domain measurement method can be used to observe the

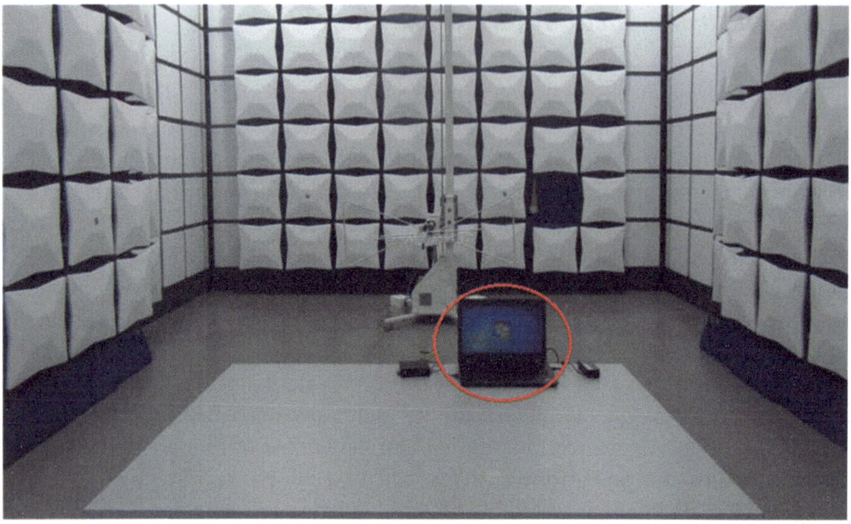

Fig. 1.1 Notebook electromagnetic radiation test environment

test target in a timely manner, to obtain the radiation spectrum, the frequency range, the radiation intensity, and other parameters.

For the test of each component of the information equipment, the method of screening isolation method, the method of the near-field detection, or injection probe can be carried out.

1.2 Conventional Methods on Electromagnetic Information Interception and Reproduction

The leakage of electromagnetic information in a computer and its information devices would inevitably lead to the risk of information interception, data reconstruction, and information reproduction of the leaked information. So it is very important for the security protection of electromagnetic information by analyzing and researching the technology of intercepting, reconstructing of data, and reproducing of information of electromagnetic information leakage. And this is also one important research area of information attack and defense technology and electromagnetic information protection technology in information warfare.

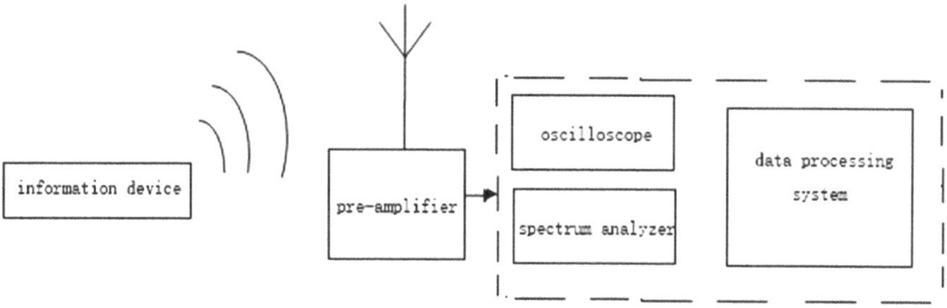

Fig. 1.2 Intercepting and reproducing device

1.2.1 Interception Purpose and Device

In view of the leakage of electromagnetic information from different parts of the information devices, a corresponding device for intercepting the leaked electromagnetic information should be adopted. Commonly used devices include antenna, near-field probe, injection probe, preamplifier, spectrum analyzer, oscilloscope, data processing system, and so on, which are shown in Fig. 1.2.

Main factors affecting the data reconstruction and information reproduction of electromagnetic information include the intercepted information amount, the sequence pulse, and the device working mechanism.

1.2.2 Key Factors of Data Reconstruction and Information Reproduction

The amount of intercepted information determines the quality of data reconstruction and information reproduction. In order to increase the amount of intercepted information, the average processing technique is an effective method, which is to get the weight average value of the intercepted information obtained in multiple sampling on the same screen, this technique can offset the lack of information through a single sampling, so as to obtain, as far as possible, the complete characteristic spectrum samples of the original information and to achieve the purpose of data reconstruction and information reproduction.

Sequence pulse is the key to interpret the intercepted information. To reproduce information of video images, the methods of extracting pixel clock synchronization signal, data-enabled synchronizing signal, video line synchronization, and frame synchronization signal are often adopted. The parameter adjustment in the line synchronization is a gradual approaching process. When the line synchronization deviation reaches a certain precision, the characteristics of the image is changed, which could be used to improve the precision of adjustment gradually. Using a special direct division circuit (DDS), a frame synchronization signal is obtained through frequency division of the received line synchronization signal so as to obtain a pixel clock synchronization signal and a data-enabled synchronization signal. When these synchronization signals are obtained, the data reconstruction and information reproduction for video could be realized.

There are differences in working mechanisms (data formats, communication protocols) for different types or models of devices. Under the condition that the amount of the information intercepted and the sequence pulse meet the requirements, it is still necessary to understand the working mechanism of the specific device to reconstitute and reproduce the data interpreted. For example, in the data reconstruction of the PS/2 keyboard, it is only possible to interpret the key scan code from the intercepted information according to the sequence pulse and the PS/2 communication protocol.

1.3 Electromagnetic Information Leakage Protection

Aiming at the radiation characteristic of the information devices, the electromagnetic information leakage protection technology is used to achieve the goal of "anti-leakage", "anti-interception", and "anti-reconstruction" by means of certain techniques, which are special technologies to prevent and restrain electromagnetic information leakage. Being used as the electromagnetic information protective measures, the first consideration of this technology is anti-leakage, which is regarded as the first layer of protection measures to prevent the electromagnetic information leakage by restraining the electromagnetic information leakage within the radiation source of the information devices to realize low radiation or even no radiation as far as possible. The anti-interception is regarded as the second-layer protection measures of electromagnetic information leakage, whose purpose is to increase the difficulty of intercepting the electromagnetic information. The anti-reconstruction is regarded as the third-layer protection measures, whose key point is to increase the difficulty of the original data reconstruction and the original information reproduction from the intercepted electromagnetic information. Thus, the electromagnetic information protection can be achieved by three-layer protection countermeasures.

1.3.1 Importance of Electromagnetic Information Leakage Protection

Most of the electromagnetic leaks carry the information processed and transmitted by devices, it varies in degrees only. With the improvement of technologies relating to intercept devices for electromagnetic information, which including the detection distance, receiving sensitivity, the miniaturization of receivers, the separation of micro signal from noises, the fast and accurate recognition, and so on, the requirement for security protection technology of electromagnetic information is becoming increasingly high. The situation of electromagnetic information security is becoming more serious, the protective research and methods on related technologies are even more urgent and important.

Open reports show that the electromagnetic wave radiation of the ordinary computer display can be received and reproduced in a few hundred meters or even a kilometer distance; electromagnetic information leakage of ordinary printer, fax machine, and telephone and other information processing and transmitting equipment, can also be intercepted and reproduced within a certain distance through specific means. The technology of interception and reproduction of electromagnetic information leakage has become an important channel for a lot of national intelligence agencies, to steal confidential information from other countries.

With the widespread adoption of information devices in confidential departments, security agencies, and banking and financing field, and with the enhancement of the intelligent degree of crimes, the problem of electromagnetic information leakage will become more and more prominent. Therefore, for users using a large amount of information devices, especially computer users from confidential departments, it is very important to master and apply effective electromagnetic information protection measures.

1.3.2 Conventional Methods of Electromagnetic Information Leakage Protection

The conventional electromagnetic information protection methods include the inclusive type, the source restraining type, and the red and black separating type. Among them, the inclusive type uses all kinds of electromagnetic radiation hardware protection measures to hold and seal both the useful electromagnetic signal and the interference signal within the shielding body in order to solve the problem of electromagnetic information leakage. The source restraining type is an electromagnetic information protection technology, which adopts low radiation and synthetically optimal methods in designing of the radiation sources including the system components, connectors and so on. The red and black separating type only protects the red signal of the devices.

At present, the electromagnetic radiation protection technology mainly refers to: the shielding technology, the filtering technology, the application of low-radiation components and devices, low-radiation arrangement of wires and circuit design technology, the application of noise interference sources, the optical fiber transmitting technology, the infrared laser communication technology etc. [4].

1. **Shielding Technology**

The shielding technology is to put the information devices in the shielding room or Faraday shielding box to prevent electromagnetic radiation. Among all, this is the most direct and reliable radiation protection technology. Using shielding materials on all kinds of information devices to decrease radiation or reinforce shielding, for example, the electromagnetic information leakage protection glass is installed on the display devices, which can solve the problem of electromagnetic information leakage of the display window to some extent; the shielding mesh gasket and ventilation waveguide are installed on the interfaces, aperture or holes of the information devices. The shielding materials included the finger spring and the shielding seal strip are installed in the surroundings of doors and door frames to solve the problem of electromagnetic information leakage through aperture and holes. The shielding materials included the electromagnetic information leakage protection glass, the metal finger spring, the ventilation waveguide, the mesh gasket, and others which can prevent the vast majority of electromagnetic information leakage through measures of the closed loop and ground connection and or so, which increases greatly the difficulty of interception. It is impossible to restore information clearly and completely for the purpose of confidentiality.

2. **Filtering Technology**

The filtering technology is a complement to the shielding technology. The shielded devices and components are not completely sealed in the shielding body, there are still power lines, signal lines, and public ground wire need to connect to the outside. As a result of such kind of connection, the electromagnetic waves can still radiate electromagnetic information through conduction or radiation, or transmit from outside into the shielding body through a variety of transmission lines. Using the filtering technology, only signals of certain frequency range are allowed to pass through, others are prevented. Thus, the filtering effect greatly reduces the radiation of electromagnetic information.

3. **Application of Low-Radiation Devices**

Low-radiation devices, also known as TEMPEST devices, are an important measure to prevent electromagnetic information radiation leakage. In designing and producing of these devices, the anti-radiation measures such as low-radiation components, parts, low-radiation wire arrangement, and circuit design have been adopted to minimize the electromagnetic leakage of the information devices. Besides, as the display is a weak link concerning information security of the devices, the stealing of information from the electromagnetic radiation of the display has already become a mature technology. For this reason, the using of low-radiation display is very important. For example, the electromagnetic radiation of a monochrome display is lower than that of a color display, a plasma display or a liquid crystal display also has a lower electromagnetic radiation.

4. **Application of Noise Interference Source**

The electromagnetic radiation interference technology is, during the operation process of computers and information devices, electromagnetic noise radiation is produced by the application of the interference sources, which is mixed and superimposed with electromagnetic radiation from computers and information devices and radiated outward, so that the electromagnetic information leaked from computers and information devices is not easy to be received and reproduced, so as to realize electromagnetic information security. Generally speaking, the electromagnetic radiation produced by the interference source should not exceed EMI standard, otherwise, it will cause electromagnetic interference to affect the normal operation of the information devices. Meanwhile, the interference source should be placed as far as possible from the operator to reduce the harm from the electromagnetic wave generated by the interference source [5].

To sum up, these main protective measures of electromagnetic information security are based on hardware electromagnetic radiation protection, although they can be effective immediately, the cost is high and the protection efficiency is low. Furthermore, it is difficult to embed these measures in the current system, and the later stage maintenance cost of these measures is relatively high.

1.3.3 New Type of Methods for Electromagnetic Information Leakage Protection

At present, the new type of methods for electromagnetic radiation protection mainly include: the pixel clock randomization, RGB color configuration, the image with noise, the TEMPEST font of Fourier–Gauss method, etc.

1. **Pixel Clock Randomization**

When displaying the image, the display needs control signals including horizontal synchronized signal, vertical synchronized signal, data-enabled signal, and others, all these synchronous control signals are set up based on the pixel clock. Usually, the pixel clock in a device is a fixed value. Once the pixel clock changes, it will affect the numerous control signals mentioned above. This feature can be used for information security protection.

2. **RGB Color Configuration**

RGB color configuration technology can optimize the configuration of button color to make the D-value of the relative analog voltage to be kept constant in key operation process, which is the constant D-value of voltage analog of RGB signals of adjacent pixels, so as to prevent electromagnetic information from leaking. This technology is widely used in devices requiring the input of validating code or password, like the touch screen devices.

3. **Image with Noise**

Image added with noise is a means of electromagnetic information protection, which randomizes electromagnetic leakage signal in order to reduce the signal-to-noise ratio of the leaked signal. The purpose is to prevent the recovering of effective image information from the information intercepted. This technology embodies the anti-reconstruction strategy of electromagnetic leakage protection.

4. **TEMPEST Fonts based on Fourier-Gauss Method**

TEMPEST font of Fourier–Gauss technology is an improvement of traditional TEMPEST font technology which perform Gaussian filtering again in addition to Fourier transformation, its main function is to make the image smooth, keep high correlation of adjacent pixels, this technology has overcome the "edge jitter" of the traditional TEMPEST font and the defects that the electromagnetic spectrum of the graphic character is easier to be distinguished from noise character, it has greatly improved the traditional TEMPEST font technology and is a more effective and reliable electromagnetic information software protection technology.

1.3.4 Protection Materials of Electromagnetic Information Leakage

Electromagnetic information protection materials are of various kinds and types, in addition to some conventional materials, a variety of new materials, such as the intelligent, composite electromagnetic shielding materials are constantly emerging. Especially, there are some nonconductive medium materials, owing to its transmission of nonelectrical signals in the communication process, there is no such problem as electromagnetic radiation leakage, and these materials will play a revolutionary role in promoting the technology of electromagnetic information protection.

1.4 Current TEMPEST Standard Research Status and Development Survey

Today, the research on the field of electromagnetic leakage, testing, and protection focuses on the research of TEMPEST technology. TEMPEST technology is a series of technologies including analyzing, testing, intercepting, reconstructing, reproducing, and protecting of the sensitive information carried by the electromagnetic leakage signals.

1.4.1 Analysis of Relevant Test Standards on Electromagnetic Radiation

The phenomenon of electromagnetic radiation caused by the operation of information devices is an objective existence, and this phenomenon will bring two problems, the electromagnetic interference and the electromagnetic information leakage. To solve these two problems, standards organizations at home and abroad have successively set up corresponding technical standards, EMC standards and TEMPEST standards.

1. **EMC Standard for "Anti-Interference"**

EMC (Electromagnetic Compatibility) standard mainly studies and solves and guarantees the electrical and electronic equipment to work normally, and not to produce electromagnetic interference unbearable to other equipment or human body. Owing to a strong correlation between EMC standard and TEMPEST standard, TEMPEST standard is analyzed compared with EMC standard, which has become one of the effective ways of TEMPEST study.

2. **TEMPEST Standard for "Anti-Leakage"**

TEMPEST standard is aimed to solve the electromagnetic leakage problems, and is proposed first by the United States. There are the corresponding TEMPEST standard series in Britain, Germany, France, Russia, and some other Western developed countries. China also put forward its own TEMPEST standard in the 90s of the twentieth century.

As the construction of China's TEMPEST standard started late, although, after years of development, some achievements have been made, there are still much to do to be in content with the international level of information stealing and protecting. And the development of the standard does not meet with the high-tech development level, which needs to improve constantly.

1.4.2 Research and Development from Abroad

The United States is one of the earliest countries to develop the anti-electromagnetic radiation leakage (TEMPEST) computers. It has accumulated a history of several decades for such technology, and has drawn up a series of standards from technology to management. Now, the United States has built a strict and perfect system for the secrecy and management of TEMPEST technology. TEMPEST counter working group (TCMWG) headed by the US National Security Agency (NSA) is the organization structure of such system. Next, we will take the United States as an example to describe the general situation of research and development of foreign TEMPEST technology.

1. **Constant Update on TEMPEST Standards and Specifications**

The United States began to draw up TEMPEST technical standards from the 1950s, so far, a complete set of TEMPEST technical standards has been formulated, and versions have been updated many times [6]. The update is made to adapt to higher requirements of technology development and information security, and the aim of the update is to avoid the situation of both "the lack of protection" (resulting in the leakage of confidential information) and "excessive protection" (resulting in the waste of people, money, and materials). Meeting the premise of information security, TEMPEST equipment of different grades should be chosen according to different needs.

2. **Continuous Development on Various New TEMPEST Computers**

From 1970 to 1982, there was the first-generation TEMPEST computer, also known as an inclusive computer. Based on the principle of shielding, filtering, and absorbing, the first generation of TEMPEST computer has the main body of the metal structure. It is the computer of shielding structure that limits the electromagnetic energy emission. From 1982 to 1996, there was the second-generation TEMPEST computer, also known as the red and black separating TEMPEST computer, which is based on the principle of separating red signal from black signal, and controls the emission of red signal. Since 1997, there was the third-generation TEMPEST computer, also known as Soft-TEMPEST computer, which is based on the Soft-TEMPEST technical theory, it uses software to control the red signal emission, and a dedicated attack program has been added to it as well.

3. **Increasingly Strict on Management of TEMPEST Market**

TEMPEST market in the United States reached prosperity as never before in the 1980s, but the product quality was hard to control because the quantity of the products is too much. In order to improve the quality of TEMPEST products, the United States began to implement TEMPEST product registration institution from 1989, such institute put forward strict quality requirements for TEMPEST products and established TEMPEST product certification program, under which qualified products were registered in ETPL. This plays an important role in improving the quality of TEMPEST products and preventing the producing and selling of inferior products.

1.4.3 Domestic Research and Development Survey

In view of the importance of TEMPEST technology, national plans and projects like National 863 Plan and State Planning Commission special projects have all listed the electromagnetic leakage emission protection technology into the scope of information security theme, investment in research and development as well as in popularization and application of technologies in this field has been increased. By 2010, the total sales of China's information security market were over 8 billion RMB yuan, but the proportion was obviously lower than that of the United States.

The study of TEMPEST in China started in the mid-1980s. In the early 1990s, the study of key issues of TEMPEST technology has been finished by some major achievements, including computer electromagnetic leakage and the protection mechanism, interception and reproduction technology of micro-computer radiation information, security assessment, technical product evaluation, laboratory and field tests, red and black signal recognition, electromagnetic leakage protection technology of micro-computer system and so on:

(1) As for TEMPEST technical standards, a series of national security standards have been initially formed;

(2) In the aspect of TEMPEST protection, protection products in different degrees and levels have been manufactured included metal shielding materials, filter, display shielding windows, aperture sealing materials, shielded cable, cutoff waveguide vent and shielded connector, shielding room, low leakage protection products, and electromagnetic interference products, and so on;
(3) In TEMPEST testing area, we can already measure the spurious signal leakage in the range of 18 GHz–1 kHz, and can make a quantitative analysis of the leaked signal, confirm the quantity and degree of the leakage, determine whether the products meet the TEMPEST standard.

Although attention has been paid to and study has been made of TEMPEST for many years within the country, the degree and range of its threat has not got enough understanding, recognition of contradictions between TEMPEST attack and defense is relatively one-sided, there is still a large gap, compared with that of foreign advanced countries, in TEMPEST technology, protection products, test methods, and standards. And with the development of information technology, TEMPEST must be confronted with new circumstances and new subjects. Therefore, enough attention should be paid on the research of TEMPEST threat and the corresponding detecting and evaluating technology to ensure the information security of our country.

Chapter 2
Electromagnetic Information Fundamental and Leakage Mechanism

On the basis of the theory of electromagnetic radiation and information propagation, this chapter describes the generation of electromagnetic wave and mechanism of electromagnetic radiation leakage, and discusses the ways of electromagnetic information leakage of information devices and corresponding anti-leakage methods. In view of the internal relations and its ways of expression between the digital information and the electromagnetic radiation from the information devices, combining the principle of information theory, the electromagnetic information model is established; meanwhile, a model for reconstruction and reproduction under noise condition is given to answer the question whether the electromagnetic information intercepted would be able to be reconstructed and reproduced.

2.1 Electromagnetic Information Fundamental

The information devices will inevitably produce electromagnetic radiation leakage in the process of powered work; the radiated electromagnetic signal reflects the operative mode of the current circuit, which, to a certain extent, is correlated with the information processed by the device. Would these electromagnetic signals be intercepted? It is possible to reconstruct and reproduce the relevant information through analyzing and processing, thus endangering the information security of the device. This indicates that the electromagnetic information leaked from the information device carries information. Before analyzing and solving the leaked electromagnetic information, it is necessary to understand the principle of electromagnetic radiation.

2.1.1 Principle of Electromagnetic Radiation

By electromagnetic theory, electric charges in space generate electric fields around it; the moving charges create electric currents, in addition to the electric fields, they also

T. Liu and Y. Li, *Electromagnetic Information Leakage and Countermeasure Technique*,
https://doi.org/10.1007/978-981-10-4352-9_2

produce magnetic fields in their surroundings. Electromagnetic waves are formed by the electric field and the magnetic field in space in forms of waves with the features of same phase oscillations and perpendicular to each other. The direction of propagation is perpendicular to the component plane of the electric field and magnetic field, which can effectively transfer energy and momentum [7–9].

1. Maxwell Equation

Maxwell's equations are the core of electromagnetic theory and the basic equations describing macroscopic electromagnetic phenomena. It is the theoretical foundation and the starting point on research of macroscopic physical phenomena, which can explain optical, electric, and magnetic phenomena and reveal the incentive rule between electromagnetic field generated by the current and the charge, and the electric field and the magnetic field each other, and reflect the universal law of macroscopic electromagnetic phenomena.

The basic Maxwell equations include integral forms and differential forms. The integral forms of Maxwell equation are

$$\oint_{l} E \cdot dl = -\iint_{S} \frac{\partial B}{\partial t} \cdot dS \tag{2.1}$$

$$\oint_{l} H \cdot dl = \iint_{S} \left(J + \frac{\partial D}{\partial t}\right) \cdot dS \tag{2.2}$$

$$\oint\!\!\!\oint_{S} D \cdot dS = \iiint_{V} \rho dV \tag{2.3}$$

$$\oint\!\!\!\oint_{S} B \cdot dS = 0 \tag{2.4}$$

The differential forms of Maxwell equation can be obtained from their integral forms. The differential forms of the Maxwell equation are

$$\nabla \times E = -\frac{\partial B}{\partial t} \tag{2.5}$$

$$\nabla \times H = J + \frac{\partial D}{\partial t} \tag{2.6}$$

$$\nabla \cdot D = \rho \tag{2.7}$$

$$\nabla \cdot B = 0 \tag{2.8}$$

In the above formulas, E represents the electric field intensity (V/m), H represents the magnetic field intensity (A/m), D represents the electric displacement vector/electric flux density (C/m^2), B represents the magnetic induction intensity/magnetic flux density (Wb/m^2), J represents the current density (A/m^2), ρ represents charge density (C/m^3).

Formula 2.1 refers to the electromagnetic induction theorem for Faraday, which shows that the electric field intensity E's circulation to any closed path depends on the change rate of magnetic induction intensity $\frac{\partial B}{\partial t}$'s enclosing area of this closed path. This reflects that the changing magnetic field can produce electric field. Its differential form shown in Formula 2.5 indicates that the changing magnetic field is the vortex center of the vortex field.

Formula 2.2 is Ampere theorem, and it shows that the magnetic field intensity H's circulation to any closed path depends on conduction current density J, which passes through the enclosing area of this closed path, and the flux $\frac{\partial D}{\partial t}$ of the change rate of electric field intensity. This reflects that the changing electric field must be accompanied by the magnetic field. Its differential form shown in Formula 2.6 indicates that both the conduction current density and the changing electric field are the vortex centers of the vortex magnetic field. In other words, both the electric current and the time-varying electric field can stimulate the magnetic field, and they are the vortex source of the magnetic field.

Formula 2.3 is Gauss theorem of electric field, which reflects that the charge stimulates the electric field in a divergent way. Its differential form in Formula 2.7 indicates that the charge density is the divergence source of the electric field.

Formulas 2.4 and 2.8 are the principle of magnetic flux continuity, which reflects that the fixed flux of magnetic induction intensity to any closed curve surface is zero. It represents the continuity of magnetic flux, which means the magnetic field line has neither a starting point nor an ending point. This means, in the space, there is no free magnetic charge which corresponds to the electric charge and the magnetic field can be stimulated. And, there is no divergence source for the magnetic field.

If the charge or current in the circuit changes with time, according to Maxwell's equation, then it will produce time-varying electric field and magnetic field, and they can convert each other and propagate in space at speed of light. This electric field and magnetic field propagating in space in the form of waves is called electromagnetic wave. The propagation of electromagnetic waves is also the propagation of energy, also known as electromagnetic radiation.

2. Electric Dipole and Magnetic Dipole

A device that can effectively radiate or receive electromagnetic waves is called an antenna. In the analysis of electromagnetic radiation, the antenna can be equivalent to the basic radiation element, including electric dipole and magnetic dipole, and the two are dual. The electromagnetic radiation mechanism can be described by the radiation model of electric dipole and magnetic dipole.

(1) Radiation Model of Electric Dipole

(1) Electric Dipole. An electric dipole is an infinitesimal element consisting of the length of an electric current. Any antenna loaded with a nonuniform current can

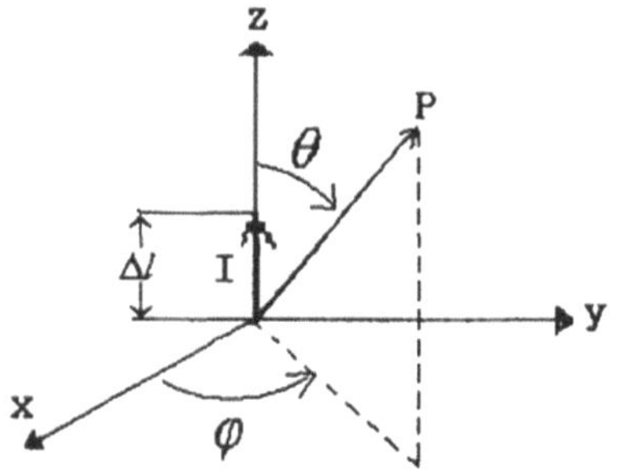

Fig. 2.1 Electric dipole model

be decomposed into innumerable infinitesimal element, and the currents on each infinitesimal element can be approximated as equal amplitude and phase, thus each element can be considered as an electric dipole, shown as Fig. 2.1.

For example, a short lead with high-frequency current, compared with the wavelength, its diameter d can be neglected, the lead length Δl is much shorter than the operating wavelength λ ($\frac{\Delta l}{\lambda} \ll 1$), and all elements of the current lead have equal amplitude and phase, which means the current in the short lead is evenly distributed. Thus, the short lead can be equivalent to an electric dipole.

The current on an electric dipole is assumed to be sinusoidal current, although time-varying current is not always changing according to sine law, but it could be decomposed into a superposition of sinusoidal currents through Fourier transform, so any time-varying current can be eventually decomposed into basic current elements.

(2) Electromagnetic Field Radiated by Electric Dipole. The electric dipole is analyzed on the origin of the spherical coordinate system. In the spherical coordinates, an observation point in the space can be represented by r, θ, φ three coordinates, where r is the length from the coordinate origin O to P (OP), θ is the angle between z-axis and OP, and φ is the angle between the projection OA of OP on x-y plane and x-axis.

The spherical coordinate system is shown in Fig. 2.2. The direction of the vector can be represented by three unit vectors r_0, θ_0, and φ_0. They represent the unit vectors for increasing direction in coordinates, r, θ, φ, respectively.

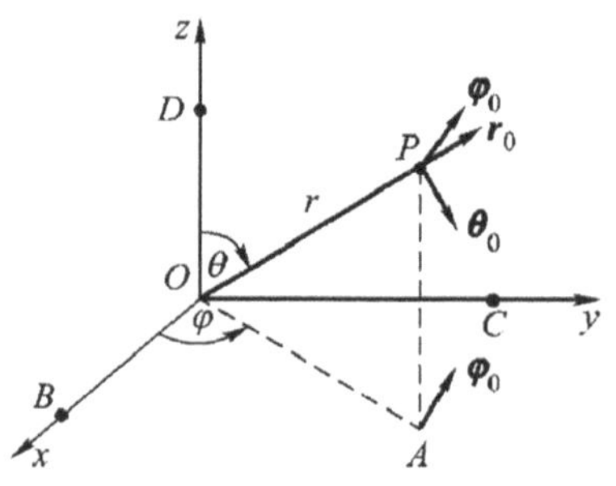

Fig. 2.2 Spherical coordinate system

To calculate combining Maxwell equations with vectors, we may find that the electric field intensity and magnetic field intensity of electric dipole radiation at point P are

$$E_r = \frac{2I\Delta lk^3}{4\pi\omega\varepsilon}\left(\frac{1}{(kr)^2} - \frac{j}{(kr)^3}\right)e^{-jkr}\cos\theta \tag{2.9}$$

$$E_\theta = \frac{I\Delta lk^3}{4\pi\omega\varepsilon}\left(\frac{j}{k\lambda} + \frac{1}{(kr)^2} - \frac{j}{(kr)^3}\right)e^{-jkr}\cos\theta \tag{2.10}$$

$$H_\varphi = \frac{I\Delta lk^2}{4\pi}\left(\frac{j}{kr} + \frac{1}{(kr)^2}\right)e^{-jkr}\sin\theta \tag{2.11}$$

In the formula, $k = 2\pi/\lambda$; λ is the wavelength, $\omega = 2\pi f$; ε is permittivity, E_r, E_θ is the electric field component in an electric dipole, and H_φ is the magnetic field component.

Thus, there is only one component in the magnetic field intensity, and there are two components in the electric field intensity E_r E_θ, the values of which vary with distance. Moreover, the propagation directions of the electric field, the magnetic field, and the electromagnetic wave produced by the electric dipole are perpendicular to each other.

In the electromagnetic environment and the interior of the information equipment, there exist the electrostatic field, magnetic field, and electromagnetic field at the same time, but with different proportions. In view of qualitative analysis, the electrostatic field and magnetic field belong to near field, they form conductive waves on conductors in an inductive manner; the electromagnetic field is far field, it produces radiation in the form of space wave. In view of quantitative analysis, the far field and near field are separated according to frequency, the critical point of distance is R = λ/2π, and a distance greater than R = λ/2π is called as far field, that is, the radiation field, composite field is in the vicinity of λ/2, near field is the place of less than λ/2, that is, the induction field.

(3) Near Field of Electric Dipole. In spherical coordinates, the region where the distance from the field point P to the origin point is much smaller than the wavelength λ is called the near region of the electric dipole. In the near region, $kr \ll 1$, $\frac{1}{kr} \ll \frac{1}{(kr)^2} \ll \frac{1}{(kr)^3}$, $e^{-jkr} \approx 1$. Then, Formulas 2.9, 2.10, and 2.11 can be rewritten as follows:

$$E_r = -\frac{2jI\Delta l\cos\theta}{4\pi\omega\varepsilon r^3} \tag{2.12}$$

$$E_\theta = -\frac{jI\Delta l\sin\theta}{4\pi\omega\varepsilon r^3} \tag{2.13}$$

$$H_\varphi = -\frac{I\Delta l\cos\theta}{4r^2} \tag{2.14}$$

We may find in formulas above, the electromagnetic field generated by electric dipole in near region and the electric field generated by electric dipole in electrostatic field, and the magnetic field generated by current element $I \cdot \Delta l$ in a steady magnetic field have consistent expressions, so the near field was called quasi-inductive field.

In the near field, the characteristic of wave impedance and the radiation source is related to its distance from the radiation source, if the radiation source shows the characteristic of magnetic dipole with heavy current and low voltage, then the near field is mainly magnetic field; on the contrary, if the radiation source has the characteristic of electric dipole, with low current and high voltage, the near field is mainly the electric field. When the wave impedance of the radiation source is comparatively high, the electric field in the near field decays according to the law of $1/r^3$ (r is the distance from the radiation source), while the magnetic field decays according to the law of $1/r^2$. When the radiation source is low impedance, the magnetic field in the near field decays according to the law of $1/r^3$, while the electric field decays according to the law of $1/r^2$.

(4) Far Field of Electric Dipole. The so-called far field means that the distance from the source point to the field point r is far greater than the wavelength, that is, $kr \gg 1, \frac{1}{kr} \gg \frac{1}{(kr)^2} \gg \frac{1}{(kr)^3}$. Then, the electromagnetic field expressions in the far field are changed to

$$E = \theta_0 \sqrt{\frac{\mu}{\varepsilon}} \frac{jkI\Delta l e^{-jkr} \sin\theta}{4\pi r} \tag{2.15}$$

$$H = \varphi_0 \frac{jkI\Delta l e^{-jkr} \sin\theta}{4\pi r} \tag{2.16}$$

In above formula, $\sqrt{\frac{\mu}{\varepsilon}} = \eta$ is the intrinsic impedance of the medium. In the atmosphere, $\eta \approx \eta_0 = \sqrt{\frac{\mu}{\varepsilon}} = 377\ \Omega$.

By Formula 2.16, the electric field only has one component θ, if $\theta = \pi/2$, $|E|$ gets its maximum value, if $\theta = 0$, $|E| = 0$.

In the far field, no matter the radiation source is high or low impedance, and no matter the electric dipole field or the magnetic dipole field, the wave impedance is a constant value, that is, the wave impedance in free space is 377 Ω. In the far field, the intensity of the radiation field is inversely proportional to the distance, which is to decay according to the law of $1/r$.

Figure 2.3 shows the direction pattern of radiation field of electric dipole. Combining with the spherical coordinate system of Fig. 2.3, we may find that the electric dipole antenna has no directionality in φ direction. In the plane where z-axis is located, electric moment P has no radiation in its own direction, and the strongest radiation is in the direction perpendicular to the electric moment P.

In free space, the time-averaged power flow of the electric dipole radiation is

$$\langle S \rangle = r_0 \frac{1}{2\eta_0} |E_\theta|^2 \tag{2.17}$$

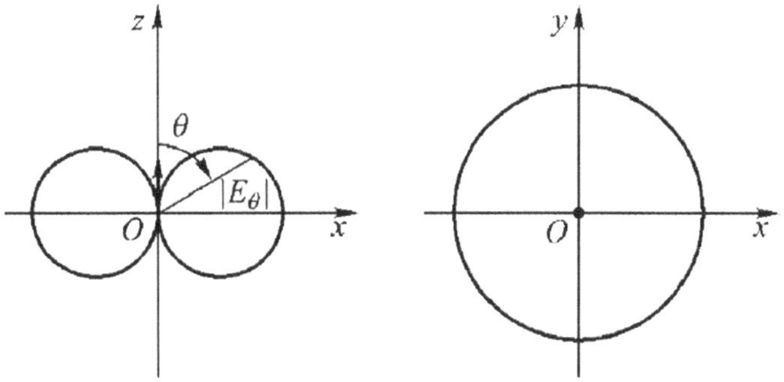

(a) x-z plane (b) x-y plane

Fig. 2.3 Direction pattern of radiation field of electric dipole

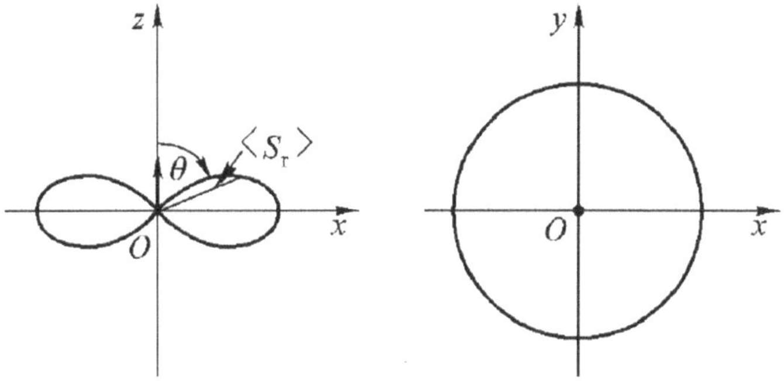

(a) x-z plane (b) x-y plane

Fig. 2.4 Direction pattern of radiation power of electric dipole

Figure 2.4 shows the direction pattern of radiation of $\langle S\rangle$ in space.

(2) Magnetic Dipole Radiation Model

(1) Magnetic Dipole. A magnetic dipole is a planar current loop with the infinitesimal radius of time harmonic current ($i = e^{j\omega t}$). Any current carrying thin wire loop L can be considered as a magnetic dipole as shown in Fig. 2.5.

(2) Electromagnetic Fields Radiated by Magnetic Dipoles. A magnetic dipole is put into a spherical coordinate system, and the magnetic field intensity and electric field intensity of the magnetic dipole radiation are obtained through Max equation and vector calculation as given below:

$$E_\varphi = -\frac{jI\Delta S\omega\mu k^2}{4\pi}\left(\frac{j}{k} + \frac{1}{(kr)^2}\right)e^{-jkr}\sin\theta \tag{2.18}$$

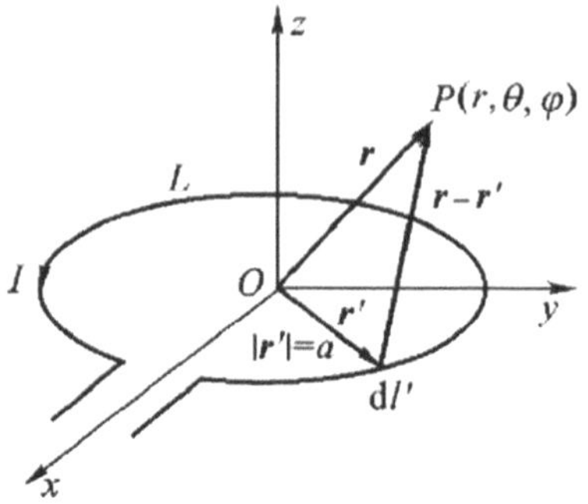

Fig. 2.5 Magnetic dipole model

$$H_r = \frac{jI\Delta Sk^3}{2\pi}\left(\frac{1}{(kr)^2} - \frac{j}{(kr)^3}\right)e^{-jkr}\cos\theta \tag{2.19}$$

$$H_\theta = \frac{jI\Delta Sk^3}{4\pi}\left(\frac{j}{kr} + \frac{1}{(kr)^2} - \frac{j}{(kr)^3}\right)e^{-jkr}\sin\theta \tag{2.20}$$

In above formula, ΔS is the area of the current loop, $\Delta S = \pi a^2$, a is the radius of the current ring.

Comparing Formulas 2.9, 2.10, 2.11 and 2.18, 2.19, 2.20, we can find that there is a dual relation between electric dipole and magnetic dipole, which reflects the duality of electromagnetic.

For information equipment, the radiation leakage units are formed by the components, parts, and connecting wires generated electromagnetic radiation, which can be equivalent to electric dipoles and magnetic dipoles and electromagnetic signal is radiated to the space around as an antenna. The leakage of electromagnetic information will be caused by the radiated electromagnetic signal with information.

The electromagnetic radiation and propagation theory, whose core is centered on Maxwell equation and the antenna model, explain the mechanism of electromagnetic leakage. But, it only depicts the objective physical phenomena. Although the electromagnetic information leakage generated by information devices is presented as physical phenomena of the electromagnetic field, it has involved the nature of digital information processed by devices. Therefore, it is not enough to explain the electromagnetic information leakage by the electromagnetic theory only in physics perspective, and we must introduce information theory to focus on analyzing and researching of the following questions:

(1) Inherent relation between the digital information processed by information devices and electromagnetic radiation.
(2) Leakage model of electromagnetic information.
(3) Necessary conditions of electromagnetic information data reconstruction and information reproduction.

2.1.2 *Electromagnetic Information Description*

In Sect. 2.1.1, the mechanism of electromagnetic leakage is expounded based on electromagnetic radiation and propagation theories, such as Maxwell equation. However, three questions are remaining to be analyzed. In the field of electromagnetic information leakage and protection, the information theories are introduced to describe the electromagnetic information and its transmission model, which have become a new and effective way to answer the above questions.

1. Electromagnetic Information Concepts

Electromagnetic information is the digital information loaded and characterized by electromagnetic wave. Electromagnetic is its external form, however, the information is its internal essence. Therefore, we must comprehend and analyze the nature and processing methods of electromagnetic information from the angle of information theories. This plays an important role in establishing the electromagnetic information model.

2. Information Transmission Model

For information devices, whether transmission of electromagnetic information to target devices is intentional or unintentional leakage of electromagnetic information, which is intercepted by monitoring devices, the above process can be abstracted as the transmission of information from the transmitting terminal to the receiving terminal. Its transmission process can be described by an information transmission model, which is shown in Fig. 2.6.

(1) Information source. Information source is the source of information.
(2) Information channel. Information channel is the message transmission channel, the intermediary from information source to information sink. It is a physical channel made up of physical medium and a logical channel for logical connections also. The medium of the information channel can be wired or wireless. During the actual transmission of information, the channel will be inevitably influenced by noise.
(3) Coding. During the processing of information transmission, signals are usually expressed by certain forms. The calculation of transforming information into signals is called coding.
(4) Decoding. The signals received need to be restored to information. The decoding work is to reproduce the message in the signals, decoding is the inverse operation of coding.
(5) Information sink. Information sink is the ending point or destination of information transmission.

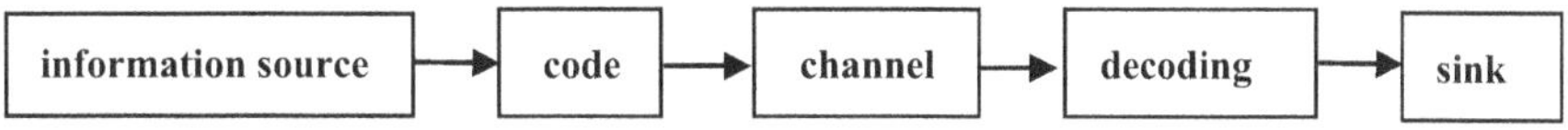

Fig. 2.6 Information transmission model

In information theories, channel capacity is usually used to measure the transmitted information. In the study of electromagnetic information transmission, channel capacity is a key factor in measuring the leakage and intercepting and reproducing the electromagnetic information.

3. Channel Capacity

According to the information theory [10], for the ideal noiseless channel, the channel capacity can be calculated by Nyquist formula.

$$C = 2B \times \log_2 M \tag{2.21}$$

In above formula, C is the channel capacity, also known as data transmission rate (b/s), B is bandwidth (Hz), M is the signal coding series, which determines code element of signals.

Owing to the interference among the code elements, the code element transmission rate in Nyquist formula is limited, and cannot be increased randomly. Otherwise, the receiving terminal cannot determine the code element correctly. Nyquist formula is derived under ideal conditions. Under real conditions, the maximum transmission rate of code element is less than the value obtained under ideal conditions.

For imperfect noisy transmission channel, there are Gaussian noise interference signals of limited bandwidth, and Shannon formula can be used to calculate the channel capacity.

Assuming additive white Gaussian noise power of the continuous channel is N (unit: W), the channel bandwidth is B (unit: Hz), the signal power is S (unit: W), then the channel capacity of such signal channel is

$$C = B \times \log_2\left(1 + \frac{S}{N}\right) \text{ (b/s)} \tag{2.22}$$

In above formula, C is channel capacity, S/N is signal-to-noise ratio, S is average signal power, and N is average noise power.

Shannon formula shows the theoretical transmission limit value of the amount of information transmitted in unit time for a transmission channel with a certain frequency bandwidth B, when the signal and the average power of the noise acting on the channel are given. There is loss, delay, and noise on a real transmission channel. Loss weakens the signal strength, which leads to decrease of signal-to-noise ratio (S/N). Delay causes distortion of the signal at the receiving end. Noise can bring damages to signals and generate error code. Shannon formula gives the limit of the information transmission rate, that is, given the transmission bandwidth and a certain signal-to-noise ratio, the upper limit of the information transmission rate is determined. And, the limit cannot be broken. To improve the transmission rate of information, we need to increase either the bandwidth of the transmission channel or the signal-to-noise ratio of the transmitted signal.

Since the noise power N is related to the channel bandwidth B, if the noise one-sided power spectrum density is n_0 (W/Hz), then the noise power $N = n_0 \times B$. Therefore, another form of Shannon's formula is

$$C = B \times \log_2\left(1 + \frac{S}{n_0 B}\right) \tag{2.23}$$

Through Formula 2.23, we know that the channel capacity of a continuous channel is limited by three factors, B, n_0, and S. As long as these three factors are determined, the channel capacity is determined accordingly.

For Shannon formula, we can make the following important conclusions:

(1) Suppose there is a noisy channel with a channel capacity of C to transmit information at a rate of R. If R < C, then there exists a coding technique that allows the errors received at the receiving end to be of arbitrarily small value. This means, theoretically, it is possible to send information without error at a maximum transmission rate of C. Conversely, if R > C, then it is impossible to achieve an arbitrarily small rate of error. Therefore, when the transmission rate exceeds the channel capacity, it is not possible to guarantee a reliable transmission of information.
(2) The channel capacity C can be improved by either reducing n_0 or increasing S/N. Especially, when $n_0 \to 0$, $C \to \infty$, which means that a free channel has infinite capacity.
(3) Increase in channel bandwidth B can also increase channel capacity C, but the increasing cannot be unlimited. This is because if S, n_0 are definite, then the channel capacity C is capped, this is shown in Formula 2.24 as given below:

$$\lim_{B \to \infty} C = \frac{S}{n_0} \log_2 e \approx 1.44 \frac{S}{n_0} \tag{2.24}$$

(4) When B and S/N are definite, then C is a fixed value. Maintaining of the channel capacity could be obtained through adjusting B and S/N, that is, the channel capacity can be kept constant through the interchanging of the system bandwidth and the signal-to-noise ratio. To reach a certain actual transmission rate, the reciprocity principle of Shannon formula could be used in system design to determine the appropriate bandwidth and signal-to-noise ratio.

Through the above theoretical analysis and discussion of problems, we can see that information theories possess guiding significance for the analysis of electromagnetic information. Such problems of transmission, measurement, distortion, and so on are also problems in the electromagnetic information propagation process; theories and information processing methods like the channel capacity and the information source and channel coding can also be adopted in analyzing and processing of electromagnetic information.

2.1.3 Correlation Between Electromagnetic and Information

The reason for information leakage by electromagnetic radiation is closely related to the structure of information devices itself. The vast majority of digital integrated circuits are constituted by gate circuit, non-gate circuit, exclusive or gate circuit, and such kind of CMOS logic gate circuit. The CMOS non-gate circuit is the simplest, but it has the basic characteristics of all CMOS logic gate circuits. Through the analysis of characteristics of CMOS non-gate circuit, we can resolve its electromagnetic radiation characteristic, and further infer its electromagnetic radiation characteristics of digital integrated circuits, and reveal the internal cause of electromagnetic information leakage of digital devices such as the computers.

In the working process of CMOS non-gate, there are three kinds of electric current, they are, respectively, the static leakage current, the dynamic short-circuit current, and the current formed through charging and did charging of dynamic load capacitance, the working principle is shown in Fig. 2.7. Regularly, greater current would consume more power and with stronger electromagnetic radiation. Among them, the static leakage current accounts for a small proportion, this can be neglected. The power consumption or electromagnetic radiation generated by a circuit is mainly caused by charging and discharging of a load capacitor. Due to the short duration of the dynamic short-circuit current, a spike pulse can be formed and can be easily monitored and intercepted.

In CMOS devices, which process digital signals, all operations are under the control of the clock, the rising or falling edge of the clock triggers the action of each component, which causes the changes of logic status. When the input signal 1 $\rightarrow$ 0, voltage V_{in} changes from high level down to low level, the output signals 0 $\rightarrow$ 1, voltage V_{out} converts from low level to high level, dynamic load capacitance charging current appears in the circuit; when the input signal of 0 $\rightarrow$ 1, voltage V_{in} jump from low level to high level, the output signal 1 $\rightarrow$ 0, voltage V_{out} converse from high level to low level, dynamic load capacitance discharging current appears

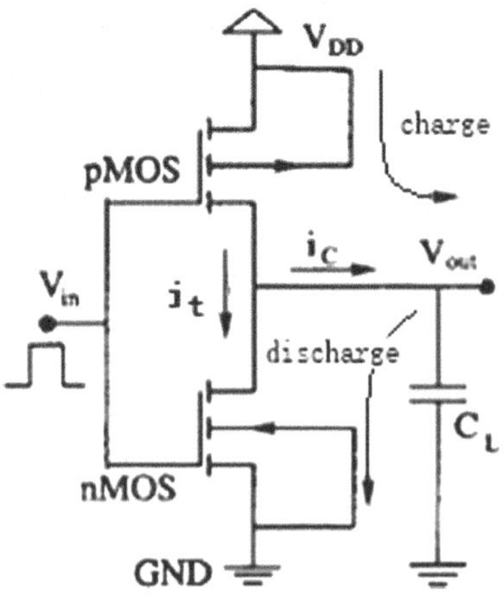

Fig. 2.7 CMOS non-gate circuit illustrative diagram

in the circuit; when the circuit works, the load capacitances charge and discharge according to different input and output signals, and this forms the charging and discharging currents i_c. At the same time, each time the input signal 1 → 0 or 0 → 1, during the transient process, pMOS and nMOS are simultaneously connected to generate the dynamic short-circuit current i_t. The short-circuit current i_t and the charging–discharging current i_c act together to form the i_a, and the information device generates electromagnetic radiation, which is shown in Fig. 2.8 [11].

In information devices, components like the monitor, ALU, memory, bus, and so on are all composed of CMOS logic gate circuit. These components process information under the control of the clock cycle and the process includes data reading, data operation, data storage, etc. Information processing is accompanied by the change of data represented by the flipping of data 0 and 1, which leads to the logic state change of the CMOS gate circuit, causing the charging and discharging of the CMOS gate circuit resulting in electromagnetic radiation, as shown in Fig. 2.9. In summary, for

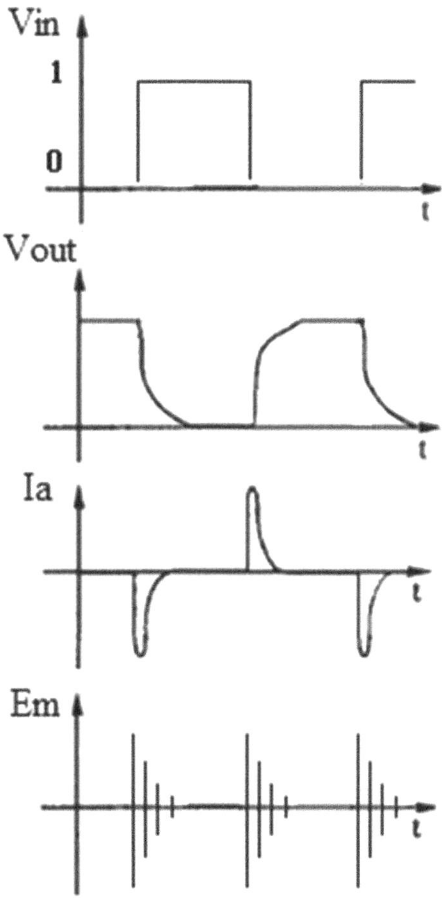

Fig. 2.8 Relations between gate status change and electromagnetic radiation

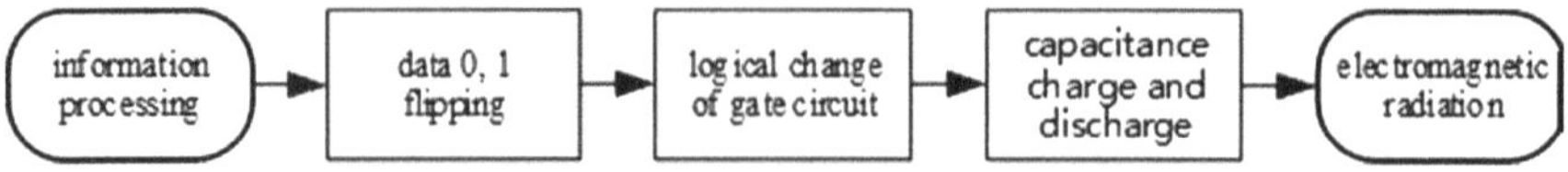

Fig. 2.9 Correlation of electromagnetism and information

information devices, there exists a correlation between the electromagnetic radiation and the information processed.

2.2 Electromagnetic Information Leakage Model

At present, the research of electromagnetic information security represented by TEMPEST, mostly aimed at the physical phenomenon of electromagnetic radiation leakage, which is based mainly on the electromagnetism-related theory, not enough theoretical explanation about the characteristics of information carried by the leaked electromagnetic waves is given, thus causing lack of systematic model depiction of the electromagnetic information leakage. In view of the correlation between the electromagnetic radiation of the information devices and its information processed, it is of a theoretical significance for research on electromagnetic information leakage and protection, which carry out research based on combination of electromagnetic propagation theory and information theory, and establish electromagnetic information model suitable for electromagnetic information expression.

2.2.1 Electromagnetic Information Leakage Model on Ideal Environment

The ideal condition refers to the situation in which electromagnetic waves radiated by the information equipment propagates in the free space. According to Maxwell equations and the antenna theory, the components, parts, and connecting wires of the information devices generating electromagnetic radiation can be equivalent to antennas. So, the electromagnetic radiation of information devices, whether it is intentional radiation or unintentional leakage, can be correspondingly regarded as equivalent antenna electromagnetic wave emission.

According to the information theories, the equivalent of antenna of electromagnetic wave transmitting can be viewed as information source of the system, and the free space of electromagnetic wave transmitting can be viewed as information channel. Neglecting the interference of noises, all electromagnetic radiation signals can be viewed as the coding of electromagnetic information.

Under the ideal condition, diagrammatic sketch of electromagnetic information leakage model is shown in Fig. 2.10.

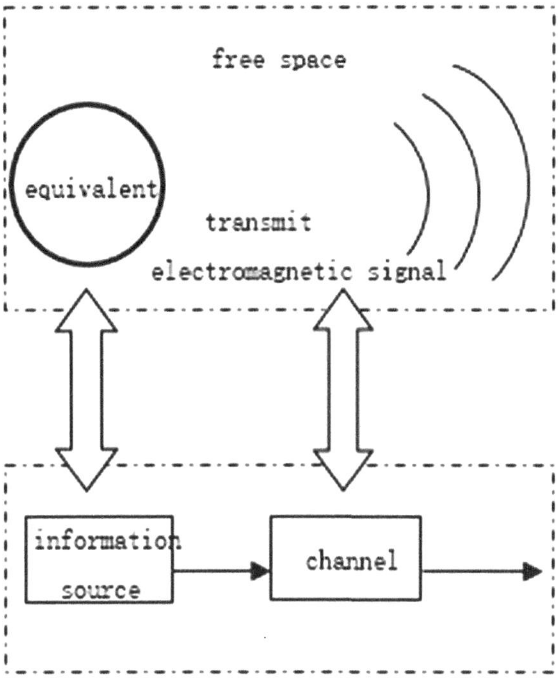

Fig. 2.10 Diagrammatic sketch of electromagnetic information leakage model under the ideal condition

Therefore, according to the Nyquist formula of information theory, the channel capacity of the information source is C

$$C = 2B\log_2 M \tag{2.25}$$

In the formula, M is the encoding series of digital signals processed by information equipment, it reflects the number of states of the digital information that can be expressed, which relates to the amount of information to be described and it influents the data transmission rate and B is the channel bandwidth. Ideally, the bandwidth of source and that of channel are equal.

When M is definite, the value of B limits the channel capacity C.

The source of information in the electromagnetic information model refers to various electromagnetic radiation sources. These radiation sources do not have a dedicated antenna. The components and wires in the device can be regarded as their equivalent antennas. The operating frequency of equivalent antenna is determined by circuit layout and material characteristics.

Nyquist formula is expressed as channel capacity of communication propagation; here, communication propagation refers to active communication mode in general sense, while electromagnetic leakage is the objective existence of unintentional radiation generated by equivalent antennas in information devices. Compared with active communication, unintentional electromagnetic leakage is limited by many objective factors. For the leakage of electromagnetic information, under ideal and noise-free conditions, the speed rate of electromagnetic information leakage depends on the channel capacity.

To establish the mathematical model of electromagnetic information leakage, the restrictive condition k_0 is introduced in Nyquist (Formula 2.25) to describe the channel capacity of electromagnetic information leakage accurately.

Therefore, the mathematical model of electromagnetic information leakage under ideal conditions is expressed in the following Formula 2.26:

$$C_{k0} = 2k_0 B \log_2 M \tag{2.26}$$

In this formula, C_{k0} is channel capacity of unintentional radiation of electromagnetic information, $\mathbf{k}_0$ is defined as the restrictive factor of electromagnetic leakage, which includes working mechanism of unintentional radiation equivalent antennas, circuit characteristics, sensitivity of receiving antenna, and others factors, so the value of $\mathbf{k}_0$ should be less than 1 ($0 < \mathbf{k}_0 < 1$).

The following factors should be considered in selecting values for K_0:

(1) Working frequency range of the information devices.
(2) Radiation intensity of electromagnetic leakage of the information devices.
(3) Maximum number of distinguishable harmonic waves to electromagnetic signals.
(4) Sensitivity of receiving antennas.

2.2.2 *Electromagnetic Information Leakage Model on Noise Environment*

Furthermore, by analyzing characteristics of random noise, we corporate it with the model described above, and discuss further necessary conditions of electromagnetic information interception and reconstruction under noise conditions, so as to establish the mathematical model accordingly.

Information channel under real conditions is always disturbed by various noises. Shannon theory discusses the issue of channel disturbed by random noises and presents the Shannon formula for calculation of channel capacity, that is,

$$C = B \log_2(1 + S/N) \tag{2.27}$$

In above formula, C represents channel capacity, W represents bandwidth, S represents signal power, and N represents noise power.

Noise has the characteristics of complex sources, wide range frequency, and strong randomness. Under conditions of adding random noises, the electromagnetic radiation, just like that under the ideal conditions, is formed according to the equivalent antenna radiation of components, parts, and connecting wires, and is also restricted by working mechanism of the equivalent antenna, characteristics of the circuit, sensitivity of the receiving antenna, and many other factors. For the above reason, the electromagnetic leakage restrictive factor $\mathbf{k}_1$ is introduced to apply for the channel capacity under the noise conditions.

Therefore, based on Shannon formula, the mathematical model expression of electromagnetic information leakage under noise conditions is

$$C_{k1} = k_1 B \log_2(1 + S/N) \tag{2.28}$$

In this formula, C_{k1} is the channel capacity of unintentional radiation of electromagnetic information under noise conditions, in which $\mathbf{k}_1$ is the electromagnetic leakage restrictive factor, and the principle and range of its value selection is the same as $\mathbf{k}_0$.

2.2.3 Electromagnetic Information Interception and Reproduction Model

Assuming that there is a channel of noise with limited bandwidth between the electromagnetic information leakage source (information source) and receiver (information destination), the model of electromagnetic information interception and reproduction is shown in Fig. 2.11.

In this diagram, the source of electromagnetic information leakage or information source refers to any information device that can cause electromagnetic information leakage. The receiver or information destination refers to any device for interception.

In order to analyze the quality of data reconstruction and information reproduction, five variables are given as follows:

I total amount of raw information processed by information devices per unit time.

C_{k0} channel capacity of electromagnetic information leakage under ideal conditions.

C_{k1} channel capacity of electromagnetic information leakage under noise conditions. C_r amount of information that can be intercepted per unit time.

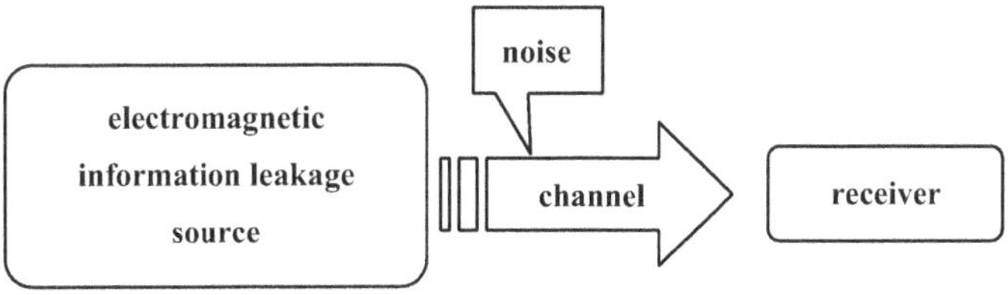

Fig. 2.11 Schematic diagram of electromagnetic information interception model under noise conditions

Q quality of data reconstruction and information reproduction.

Through analysis, we can conclude that the above variables have relations as follows:

(1) The original information processed by the information device will not be completely leaked, also the electromagnetic information leaked may not all be intercepted. On the other hand, compared with that under noise-free conditions, the channel capacity of electromagnetic information leakage under noise conditions is comparatively less.

The relation between them is

$$C_r \leq C_{k1} \leq C_{k0} \leq I \tag{2.29}$$

(2) The quality of data reconstruction and information reproduction can be expressed as

$$Q = \frac{C_r}{I} \tag{2.30}$$

When $C_r = I$, the amount of information that can be intercepted is equal to the amount of information processed by the information processing equipment. Theoretically, it is possible to realize data reconstruction and information reproduction; when $C_r < I$, the amount of information intercepted is less than the amount of information processed, and it is easy to cause the lack of information, which would reduce the possibility of data reconstruction and information reproduction.

According to literature reports, for experiments targeted at intercepting, reconstructing, and reproducing of video signals, when $Q > 0.8$ (Q denotes reproduction quality), it is able to reconstruct and reproduce the legible image, and when $Q > 0.5$, it is able to reconstruct and reproduce the basically distinguishable image, the rest of the cases would not able to reconstruct and reproduce the distinguishable image.

In fact, whether or not information can be reconstructed and reproduced depends on many factors, among which the most critical factor is the time sequence impulse. As time sequence impulse is a requirement for interpreting digital information, which can be obtained from the information intercepted, it would be possible to reconstruct and reproduce the original information; if not, there would be no possibility of reproduction and reconstruction of the original information.

2.3 Ways of Electromagnetic Information Leakage

In electromagnetic environment, the electromagnetic information leakage of information devices takes two ways: direct modulation and cross modulation [12]. In the transmission process of useful signals, there would generate the spurious modulation effect to a strong radiation periodic signal, the kind of electromagnetic radiation leakage that carries useful information is called direct modulation; the electromagnetic radiation leakage generated by a combining effect from both the useful signal and the periodic signal on the nonlinear components is called cross modulation.

There are mainly two kinds of leakage of electromagnetic information: in view of the leakage source, they are radiation emission and conduction emission; in perspective of the receiving devices, they are radiation coupling and conduction coupling.

2.3.1 Radiation Emission Leakage

Radiation emission leakage refers to the stray electromagnetic energy radiated in the forms of electromagnetic wave through the equipment shell, holes and seams in the shell, connecting cables, and so on, which is shown in Fig. 2.12. Radiation coupling is the coupling of energy in space to a receiver in the form of electromagnetic waves. The radiation coupling includes the coupling of space electromagnetic waves to receiving antennas, or to the cables and the coupling of cables to the cables, which are the main mode of confidential information leakage through transmitting and processing devices like computers, communication, network, and others.

No matter through which way the electromagnetic information leaks, the information device is an important source of leakage. Analysis of the radiation emission and conduction emission of the leakage in information devices is of great signif-

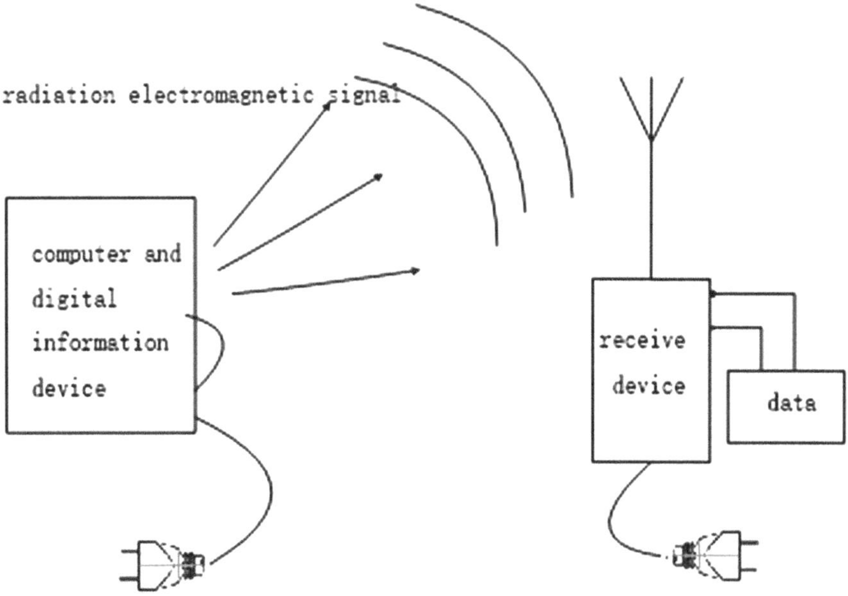

Fig. 2.12 Diagrammatic sketch of radiation emission

icance for both understanding leakage mechanism and guiding the protection of electromagnetic information.

Components, parts, and connecting wires that generate electromagnetic radiation in an information device constitute a radiation emission leakage unit, which can be viewed as being composed of various sizes of electric vibrators and current loops. Under certain conditions, it can also be viewed as being composed of electric dipoles and magnetic dipoles.

Suppose there is a current source with a length of dl in the circuit, equalize it to electric dipoles to analyze the electromagnetic field. The current flowing through the circuit is I the electromagnetic field at the point $p(r, \theta, \varphi)$ is

$$E_r = \frac{2Idlk^3}{j4\pi\omega\varepsilon}\left(\frac{j}{k} + \frac{1}{k^2}\right)\frac{e^{-jk}}{k}\cos\theta \tag{2.31}$$

$$E_\theta = \frac{Idlk^3}{j4\pi\omega\varepsilon}\left(\frac{j}{k} + \frac{1}{k^2 - 1} - 1\right)\frac{e^{-jk}}{k} \tag{2.32}$$

$$H_\varphi = \frac{Idlk^3}{j4\pi}\left(\frac{j}{k} - 1\right)\frac{e^{jk}}{k}\sin\theta \tag{2.33}$$

In the formula, $k = 2\pi/\lambda$, λ is the wavelength, $\omega = 2\pi f$, E_r, E_θ is the electric field component in an electric dipole, and H_φ is the magnetic field component.

The total radiation power of the electric dipole field is

$$P = \frac{Z_0 k^2 (Idl)^2}{12\pi} \tag{2.34}$$

In this formula, Z_0 is the impedance in vacuum, $Z_0 = 120\pi$; I is the current flowing through the electric dipole; and dl is the length of the electric dipole.

Putting the value of Z_0 into Formula 2.34, we get

$$P = 40\pi^2 I^2 (dl/\lambda)^2 \tag{2.35}$$

Equalize the total radiant power to the power absorbed by a radiant resistance, there would be

$$P = \frac{1}{2} I^2 R_r \tag{2.36}$$

$$R_r = 80\pi^2 (dl/\lambda)^2 \tag{2.37}$$

The greater the radiation resistance, the stronger its radiation power, that is, the greater the value dl/λ, the stronger the radiation power, this is the reason behind the strong power of the high-frequency radiation. To restrain radiation leakage, you may reduce radiation intensity or radiation resistance or decrease, as much as possible, the frequency of the electromagnetic wave radiation.

In information equipment, a current loop with a diameter much smaller than its wavelength can be considered as a magnetic dipole. The electromagnetic field component of a magnetic dipole at a space point $p(r, \theta, \varphi)$ is

$$H_r = \frac{2I\Delta S k^3}{4\pi}\left(\frac{j}{k} + \frac{1}{k^2}\right)\frac{e^{-jk}}{k}\cos\theta \tag{2.38}$$

$$H_\theta = \frac{I\Delta S k^3}{4\pi}\left(\frac{j}{k} + \frac{1}{k^2} - 1\right)\frac{e^{-jk}}{k}\sin\theta \tag{2.39}$$

$$H_\varphi = \frac{I\Delta S\omega k^3}{4\pi}\left(1 - \frac{j}{k}\right)\frac{e^{-jk}}{k}\sin\theta \tag{2.40}$$

In above formula, ΔS is the area of the electric current loop, $\Delta S = \pi a^2$, and a is the radius of the electric current loop.

Under the condition that there are both the magnetic field and the electric field, a magnetic dipole would generate a total radiation power of

$$P = \frac{Z_0 k^4 (I\Delta S)^2}{12\pi} \tag{2.41}$$

Putting $\Delta S = \pi a^2$ into Formula 2.41, we have

$$P = 160\pi^6 I^2 (a/\lambda)^4 \tag{2.42}$$

To equalize the total radiant power into the power absorbed by a radiation resistance, and the radiation resistance is

$$R_r = 320\pi^6 (a/\lambda)^4 \tag{2.43}$$

For electric dipoles and magnetic dipoles of the same line length, since $dl = 2a$, having the same current I, the radiation energy of a magnetic dipole is much larger than that of an electric dipole.

2.3.2 Conducted Emission Leakage

Conduction emission leakage refers to stray electromagnetic energy which is conducted out through a variety of lines (including power lines, signal lines, etc.) as shown in Fig. 2.13.

Conduction coupling means that electromagnetic energy is coupled to a receiver through a metallic conductor or lumped element, such as a capacitor, transformer, in the form of voltage or current. There are three kinds of conduction coupling, direct conduction coupling, common impedance coupling, and transfer impedance coupling.

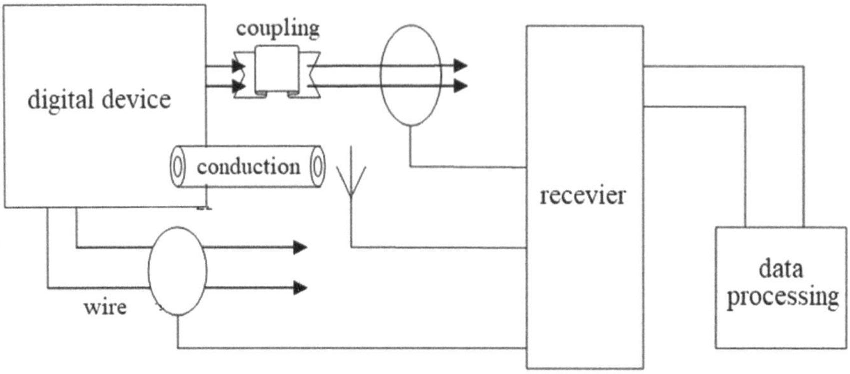

Fig. 2.13 Schematic diagram of conduction emission

1. Direct conduction coupling

Direct conduction coupling is the electromagnetic radiation impedance coupled directly through the wire, metal, resistance, inductance, capacitance, and others to the input end of receiver, thus get into the device. In addition to the coupling directly through capacitance, capacity coupling can also be generated by stray capacitance of two wire or printed circuit boards, and inductive coupling includes mutual inductance coupling between two conductors or inductive coupling aroused by common reactance of two wire or printed circuit boards.

(1) Inductive coupling. When the signal current of the leakage source reaches the remote load along the wire, mutual inductance coupling occurs on the parallel wire connected to the input end of receiver. Thus, electromagnetic information can be transmitted from the leakage source to a receiver by mutual inductance coupling, which results in electromagnetic information leakage.
(2) Capacitive coupling. When frequency is high, the coupling between two parallel wires is mainly capacitive coupling. In design of the printed circuit board and its device layout, in the section with high-frequency signals, the distance between two printed lines and the position among the devices should be considered. The closer the distance, the greater the possibility of capacitive coupling would occur. In cables of multiple wires in the radio-frequency circuit, the signal voltage on one wire can be coupled to all other wires, so the high-frequency signal lines should all be screened; in high-frequency amplifier triode, distributed coupling capacity between the pins will be produced, and for this reason, the length of the pinots should be minimized.
(3) Resistive coupling. Resistive coupling is the simplest way to conduct coupling. For example, the connecting wires of two circuits, the signal connection between devices, the power line between the power supply and the load, etc. In addition to the normal transmission of control signals and supply current, they also transmit electromagnetic radiation through wires.

2. Common impedance coupling

When there is common impedance between the output loop of the radiation source and the input loop of the receiver, a common impedance coupling will occur. Common impedance coupling mainly includes common ground impedance coupling and common power coupling. Common ground coupling means there is a common ground impedance for the signal pathways of more than two amplifiers or digital circuits on the printed circuit board of the information device, or there is public safety grounding including metal grounding wire, grounding network for more than two sets of equipment, or the connection of grounding wire to water pipe or heating radiators.
3. Transfer impedance coupling
Impedance coupling is different from direct conduction coupling in that the electromagnetic information leakage emitted by the source is not directly transmitted to the receiver, but changed by transfer impedance from radiation current into the input voltage of the receiver at the input end. No matter what kind of coupling it is, it can be described by transfer impedance coupling.

When analyzing the energy intensity of electromagnetic leakage of information equipment, the conduction emission is usually divided into two modes: differential mode conduction and common mode conduction.

(1) Differential mode conduction. Differential mode conduction is produced by the flow of signal current in the loop formed by the transmission line and its return line (signal grounding). Prediction of differential mode conduction can be made using the small loop antenna model, and then the maximum electric field intensity on the transmission direction is

$$E = 131.6 \times 10^{16} \cdot \frac{f^2 A I}{r} \tag{2.44}$$

In this formula, f is conduction current frequency, A is loop area, I is signal current intensity, and r is distance from radiation source to observation point.

Since the differential mode current is the current needed for circuit operation, its flowing direction, frequency, and intensity are known. Various measures can be adopted in design to minimize differential mode conduction. From Formula 2.44, we know that the method of restraining differential mode conduction is to reduce signal current and operating frequency. The effective way to reduce current is to select a low power circuit and add buffer properly. The effective way to reduce the frequency is to appropriately reduce the higher harmonic wave of the pulse signal, under the circumstance that it would not affect the function of the circuit. There are two methods that can be chosen, one is to choose low-speed devices and the other is to use the low-pass filter to cut off frequency properly.

(2) Common mode conduction. Common mode conduction is produced by the common mode current in the loop of the signal and the earth in the transmission line, which is not needed for circuit operation and is caused by improper design, so the prediction of it is very difficult [13]. In this case, the electric field intensity in the maximum conduction direction is

$$E = 12.6 \times 10^{-7} \cdot \frac{f^2 L I}{r} \tag{2.45}$$

In above formula, f is common mode current frequency, L is cable length, I is common mode current strength, and r is distance from the observation point to source of conduction.

Thus, an effective way to suppress the common mode transmission is to reduce the common mode current. When common mode current loop impedance is low, common mode choke can be used; but when common mode current loop impedance is higher, the common mode choke has no obvious effect, and the method of shielding and filtering needs to be used.

The electromagnetic information leakage of the information devices might occur in two ways: radiation emission and conduction emission. However, the two do not exist independently, conduction emission is inevitably accompanied by radiation emission. They are related to each other and have mutual exchange of energy. For the electromagnetic information leakage of information devices, a systematic analysis should be made considering both distinguishing characteristics of leakage paths and relations between them.

2.4 Electromagnetic Information Classification

Because electromagnetic information has characteristics of both digital information and electromagnetic wave, it can be classified according to effectiveness of the electromagnetic information and its propagation mode.

2.4.1 Electromagnetic Information Classification on Effectiveness

In processing, sending, and receiving of the electromagnetic information, often there exists not the simple single signal, but a hybrid superposition of various electromagnetic waves. Especially, in complex electromagnetic environment, electromagnetic information would also be interfered by external electromagnetic waves accompanied by electromagnetic noise. How to distinguish the effective signals from others among mixed electromagnetic signals is a problem with significance. Therefore, according to the effectiveness, the electromagnetic signals are divided into red signal and black signal.

In electromagnetic information, signal that can be reproduced into useful information after being intercepted is called red signal. Otherwise, signal that cannot be reproduced for useful information is called black signal.

In TEMPEST technology, red and black signals are important concept, and TEMPEST design can be considered as the EMC design based on the red and black concepts. Furthermore, identification and separation of red and black signals in the system are important content of TEMPEST. Among red and black signals, the leakage of the former one causes main threat to information security. The identification of red signal leakage is the key to determine whether the leakage is existing or not. For information interceptor, separation of red and black signals is the premise of obtaining useful information and reproducing the original information.

The electromagnetic radiation of red signal would induce on power and ground wires, leading to conductive radiation leakage. For unshielded black circuit, the same problem occurs, that is, the radiation on black circuit also inducing information leakage.

The red signal from electromagnetic emanation is mainly represented in two ways, one is direct modulation electromagnetic radiation and the other is cross modulation electromagnetic radiation. The former modulation refers to the red signal transmitting which generates spurious modulation effect to certain black signals with high intensity and periodicity, that is, to say, in direct modulation, black signal acts as the carrier, while the red signal performs as the modulating signal. The latter means red signal and periodic black signal as a whole radiate on nonlinear components, thereafter, whose radiation causes information leakage, and the spectral intensity of the intermodulation signal is proportional to the product of spectral intensity of the red and black signals.

2.4.2 Electromagnetic Information Classification on Transmission Means

Electromagnetic information emission can be divided into two categories, intentional transmission and unintentional leakage.

Intentional transmission refers to active process of transmitting and processing data between transmitter and receiver parties according to agreed protocol. For information sender, it is hoped that the receiver can receive and process transmitted information with no difficulty, and can even get response and feedback.

For intentional transmission of electromagnetic information, from the perspective of electronic countermeasures, it can be divided into nonantagonistic electromagnetic radiation and antagonistic electromagnetic radiation. The former refers to the radiation produced by electromagnetic activities of industrial electrical equipment, radio, television, communication, radar, navigation, and others of home parties. The latter including electronic suppression, deception jamming, directed energy attack (laser, high power microwave), and other radiation produced by electromagnetic activities of enemies. Electronic countermeasure is the key factor affecting the electromagnetic environment in information battlefield.

Unintentional leakage can be described as emission of electromagnetic information which could be intercepted and even reproduced with no subjective awareness or willingness, that is, electromagnetic emanation mentioned above is this kind of leakage. For instance, components like chip, circuit, cable, and other conductive parts generate electromagnetic field, its variation resulting in radiation with information processed. Unintentional leakage is one of the main threats to electromagnetic information security.

Chapter 3
Electromagnetic Information Interception and Reproduction

The second chapter describes the electromagnetic information interception and data reconstruction, information reproduction model, in which the source refers to the leakage source and the sink refers to interception equipment. The electromagnetic information might be leaked when information equipment processes the data, and the leaked information might be intercepted and reproduced, thus information security will be threatened. The leaked electromagnetic signal that is intercepted from the information equipment could be reconstructed to reproduce the original information, which includes signal interception, data reconstruction, and information reproduction.

3.1 Electromagnetic Information Interception

Intercepting the leaked electromagnetic signal requires using corresponding devices and through some certain means. The intercepted data is stored and processed to facilitate the data reconstruction and information reproduction.

3.1.1 Interception Aim and Methods

The interception of electromagnetic information refers to receiving the electromagnetic signal radiated through air or transmission pipeline from the target information equipment by the specific device within a certain distance [14], which is shown in Fig. 3.1. The purpose of interception is to obtain the leaked electromagnetic information, in order to reconstruct the data and reproduce the information.

Interception is similar to TEMPST testing in terms of the devices used and the applicable means, but their purpose is very different. The purpose of interception is to obtain the leaked electromagnetic information from the information equipment as complete as possible, in order to improve the quality of information reproduction.

T. Liu and Y. Li, *Electromagnetic Information Leakage and Countermeasure Technique*,
https://doi.org/10.1007/978-981-10-4352-9_3

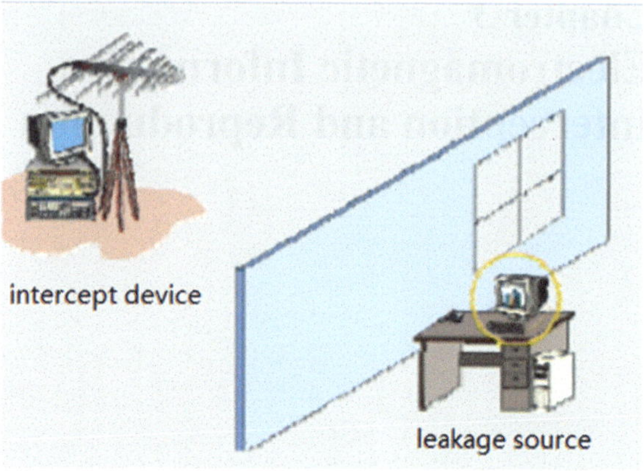

Fig. 3.1 Intercepted schematic diagram

While, the purpose of testing is to detect the electromagnetic radiation intensity of the information equipment, and determine whether it conforms to the electromagnetic information safety standard or not. In brief, the interception concerns about the full information of the target object, and the testing aims at the hazard detection of the target equipment.

The process of interception and reproduction is mainly related to electromagnetic signal acquisition, instantaneous recording, signal processing, static and dynamic database, and data mining technologies, which covers materials, electronics, antenna, test, mathematics, computer, artificial intelligence, and other professional fields, therefore, interception and reproduction is a multidisciplinary integrated technology.

Interception provides a data source for reconstruction and reproduction. Interception technology is very important, which directly affects the possibility and effect of the reproduction. To improve quality of intercepted data can be achieved by enhancing the sensitivity of the antenna, and increasing accuracy and density of data sampling.

The interception device, as shown in Fig. 3.2, is mainly consisted of an antenna system, a high-performance receiver, a preselected amplifier, a spectrum analyzer, a digital oscilloscope, and other equipment. In view of electromagnetic information leakage from different components, parts and connecting wires of information equipment, the corresponding intercepting device can be adopted, which could be an existing product, or specially customized or reformed one.

The commonly used antennas in the interception process are near-field probe, injection probe, and the various ordinary antennas, which are shown in Figs. 3.3, 3.4, 3.5, and 3.6.

For type selection, configuration, and other related information of the interception devices, please refer to Chap. 8 "Electromagnetic Information Leakage Testing" for more details.

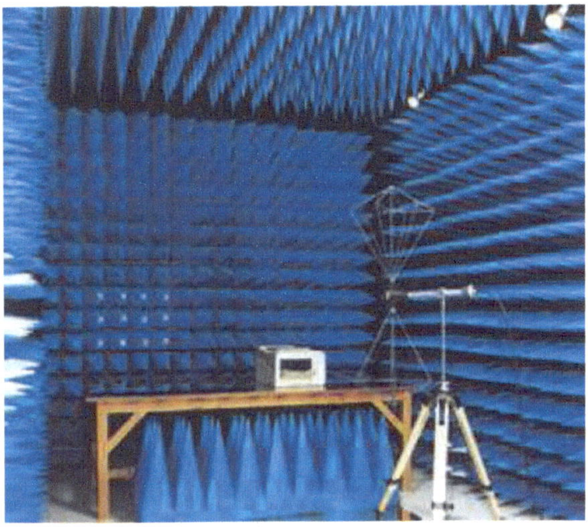

Fig. 3.2 Interception device diagram

Fig. 3.3 Injection probe

Fig. 3.4 Near-field probe

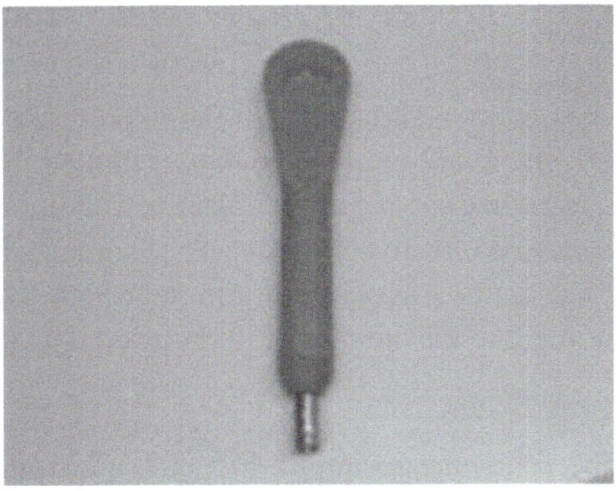

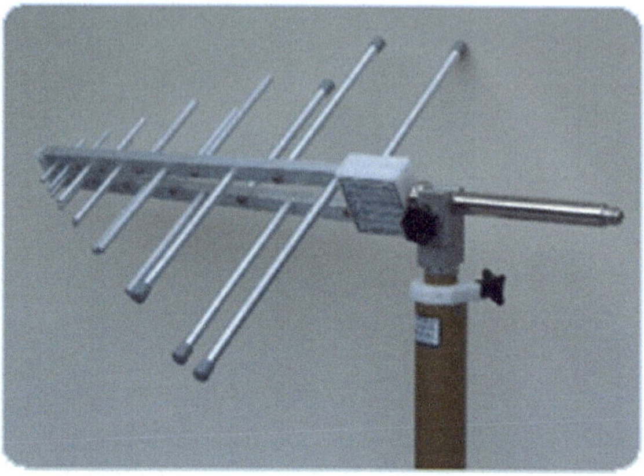

Fig. 3.5 Log-periodical antenna

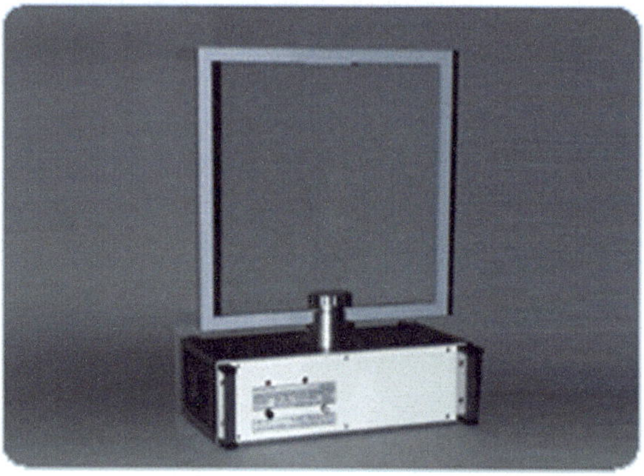

Fig. 3.6 Loop antenna

3.1.2 Process of Intercepted Data

The quality and quantity of the intercepted data are critical factors in reproduction. The amount of intercepted data is limited by both objective conditions and technical factors. In case of certain objective conditions, the amount of intercepted data can be increased effectively by means of some technical methods to improve reproduction effect as much as possible. Among which, averaging technology is a common method to improve the amount of intercepted data.

1. Averaging

The following example, Fig. 3.7, illustrates the averaging processing technique, which is about the display image interception and reconstruction from electromagnetic emanation of a laptop.

Averaging is the weighted average of the intercepted information by multi-sampling the same image, so as to compensate for the lack of information on a single sample. Therefore, to achieve the purpose of data reconstruction and information,

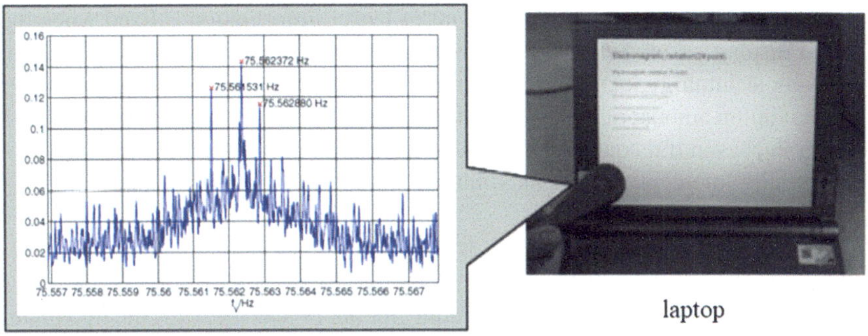

Fig. 3.7 Frequency spectrogram is leaked by electromagnetic information of a laptop

reproduction needs to obtain the characteristic samples of the original information as complete as possible. Application of the technology requires determining the number of sampling needed according to pre-estimating the maximum amount of interception.

The maximum amount of interception is usually related to the leakage of electromagnetic information radiation, while the leakage of electromagnetic information depends on corresponding information equipment. The area between the information equipment and the intercepting device can be considered as a channel, whose capacity represents the maximum amount of leaked electromagnetic radiation information passing through the information equipment in unit time. According to Chap. 2, Electromagnetic Information Leakage Model on Noise Environment, the channel capacity of electromagnetic information leakage is

$$C_{k1} = k_1 B \log_2(1 + S/N) \tag{3.1}$$

where C_r is the amount of maximum interception in unit time, it is approximately equal to C_{K1}, which represents the amount of maximum leakage of image electromagnetic radiation information.

Let I (unit: bps) be the amount of the computer video signal information, when $C_r \geq I$, enough information of a video signal can be got and a clear image could possibly be reproduced; when $C_r < I$, only some information could be obtained, so the reproduced image would be a blurred one.

The amount of leaked information generated by electromagnetic radiation cannot be greater than that of the original image, so, even if the completely leaked electromagnetic radiation signal can be intercepted, it does not reproduce all of the original information, so that averaging technology has an upper limit in reproducing the quality of the image.

For information equipment, the amount of leaked information is fixed when the image it displays is unchanged. By increasing the number of samples, the average processing technology increases the amount of interception to make it approaching to the maximum amount of leakage infinitely. When the number of samples reaches a certain value, even if it continues to increase, the times of sampling will not increase the amount of interception. The maximum times of sampling is

$$n = \lfloor M/C_r \rfloor \times f_s \tag{3.2}$$

In the above equation: M is the maximum amount of leakage, C_r is the maximum interception amount in unit time, and f_s is the sampling frequency.

The example, as shown in Fig. 3.8, shows the relationship between the original image and the image that is reproduced by the maximum amount of leakage. The upper half of Fig. 3.8 is the original black and white image, and the lower half is the intercepted and reproduced image that contains only the vertical edge information, that is, the maximum amount of leakage.

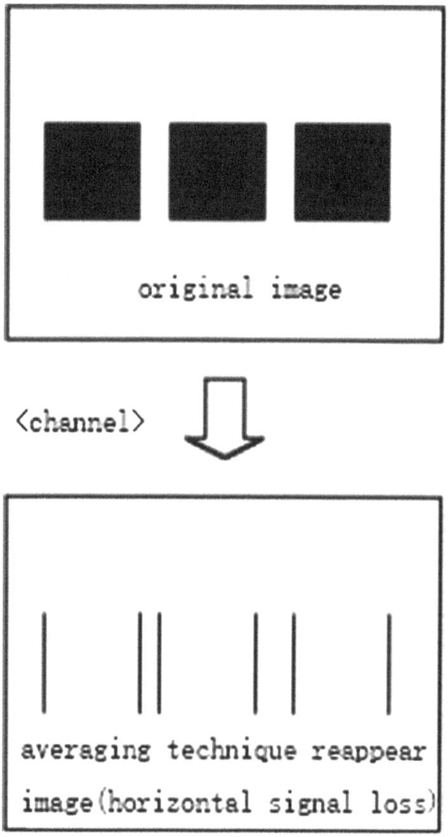

Fig. 3.8 The average processing technology reproduces the upper limited image

After N times of sampling, the amount of information intercepted by the average processing technology reaches the upper limit, only increasing the number of samples could not obtain more amount of interception, in other words, it cannot increase the characteristic information of the image so as to improve quality of the reproduced image ultimately. In Fig. 3.8, even after N times of the average sampling, the reproduced image has only a vertical line, which is the image of the color jump edge. The reason is that the information can be intercepted when the image color jumps. In addition, if the image color is constant, that is, when the accumulated value of RGB is constant, the information cannot be intercepted, thereby causing the lack of the horizontal line feature information of the image. At this point, simply increasing the sampling times N still does not help to intercept the horizontal feature information of the image.

The following is an example of reproduction using averaging technology [47]. The original displayed image as shown in Fig. 3.9, Fig. 3.10, and Fig. 3.11 show a reproduced image without using and using averaging technology, respectively.

In summary, in case of displaying the same image, application of the averaging technology can raise the total amount of the intercepted data by increasing the number of sampling times, so as to improve the quality of the reproduction image. The effectiveness of the technology mainly depends on the maximum amount of interception in unit time, while the maximum amount of leaked information determines the upper limit of its reproduction quality. To increase the number of sampling times can make the total interception amount to reach the maximum amount of leakage as much as possible, in the meantime, it also reached its upper effectiveness limit.

2. Data storage technology

For achieving data reconstruction and information reproduction, it requires sampling, preservation, and post-processing to the intercepted electromagnetic signal. For obtaining the whole electromagnetic leakage signal, a great amount of sampling data is needed, which require using the database technology to store and process, and

Fig. 3.9 The original displayed image

Fig. 3.10 No averaging technique reappear image

Fig. 3.11 Adapt averaging technique reappear image (n = 4)

setting up sample database of electromagnetic information characteristics of original information.

Database technology can effectively organize and store large amounts of sampling data, which has the advantages of reducing data redundancy, sharing data, ensuring data security, and efficiently retrieving and processing data. In addition, it can classify and store subsample data according to different targets and applications.

By using database technology, it can establish electromagnetic information characteristics sample database to record the dynamic spectrum waveform and static sampling samples of electromagnetic leakage information, so it plays the important role in the analysis and reproduction of electromagnetic leakage information:

(1) It makes ease of data access of electromagnetic leakage information. Based on the efficient access function of the database technology, it can realize secure storage backup and real-time high-speed reading and writing of massive information to ensure data security and data access timeliness.
(2) It facilitates classification management of electromagnetic leakage information that can be realized by using database table structure and association design. Data sources can be distinguished according to the category, model, and a variety of parameter characteristics of the target equipment, thus store the corresponding electromagnetic data, respectively. It can help researchers to classify, summarize, and compare various types of electromagnetic data, thus facilitates analysis and reproduction of electromagnetic leakage information.
(3) It benefits the statistical analysis of electromagnetic leakage information. Based on the electromagnetic information characteristics sample database, the data can show in the form of the frequency domain, time domain, or the combination, and displayed statically and dynamically, in charts, digital, and other types of characteristics data. Through local detailed analysis or the overall statistical analysis for data mentioned before can achieve a comparative analysis of single-group samples or multiple sets of samples.
(4) It benefits the information of electromagnetic leakage inquiry. It can provide association feature extraction and key information inquiry by setting database compound retrieval condition.
(5) It is helpful for research of reconstruction and reproduction algorithm for electromagnetic leakage information. It can classify, compare, decompose, calculate and convert the electromagnetic data by using the database and its management system according to the time domain, frequency domain, amplitude, phase, and waveform of the electromagnetic signal characteristics.
(6) It can provide the data support for electromagnetic leakage information detection. As the fundamental data support, the electromagnetic information characteristic sample database can supply the data service of storage, inquiry, update, and reading to electromagnetic leakage information detection and analysis system.
(7) It helps to judge the tested target equipment. After intercepting the electromagnetic leakage information of the target equipment, the model or model range of the target equipment and its main components can be judged by comparing with the electromagnetic information characteristic samples stored in the database. On the other hand, if model of the target equipment is known, information of other relevant parameters can be obtained.

3.2 Electromagnetic Information Reconstruction and Reproduction

Processing the electromagnetic intercepted information to realize data reconstruction and information reproduction is closely related to three elements, which are the total amount of original information, the amount of leaked information, that is the channel capacity of the electromagnetic leaked information, and the amount of intercepted information processed by the information equipment.

In Chap. 2, the concept of quality Q used in data reconstruction and information reproduction has been introduced. When the ratio of number of interception information and total amount of original information reaches a certain threshold, it is possible to reconstruct the data and reproduce the original information clearly. The clock pulse plays a key role in transmission and reading of data in the digital circuit for data reconstruction and information reproduction, so the clock pulse is an integral part in electromagnetic information interception. Furthermore, to choose the corresponding reconstruction technique for reproduction of original information requires understanding the working mechanism and data transmission format of the specific equipment [15].

3.2.1 Data Reconstruction and Information Reproduction

The aim of information interception is data reconstruction and information reproduction; whose purpose is to restore the original information as much as possible. However, data reconstruction and information reproduction are not exactly the same, they are both different and complementary.

Under certain rules, structuring data in order and expressing a clear meaning can form information. In that sense, data reconstruction refers to analysis and reorganizing the intercepted signals, while information reproduction is a form of presenting the original information.

Data reconstruction requires understanding the working mechanism and data format of the source leakage information equipment to parse and reorganize the data in sequence. For example, reconstruction of the intercepted information from keyboard needs to understand its types and protocols, after data reconstruction, the key information of that keyboard can be obtained to help for understanding the original information, but it does not involve information reproduction.

The data reconstructed can be used to reproduce the information, and quality of the reproduction information is related to the amount of reconstructed data. For instance, reproducing information of the displayed image requires reconstructing the intercepted signal and then reproduces the image information. Whether the reproduced image is clear or not depends on the amount of captured reconstruction data reaches a certain proportion of the total amount of information or not.

3.2.2 The Important of Clock Pulse

Clock pulse is the red signal of information equipment. It is required for all digital signal-processing circuits to have a clock pulse, which plays a critical role in the digital circuit to realize its function [16]. As the signal formed by the sequential circuit is mostly periodic rectangular waves, its harmonic frequency components are discrete and times of fundamental frequency, therefore, the radiation frequency spectrum of the sequential circuit is also discrete. The digital information signals are generally nonperiodic signals, and its radiation frequency spectrum is a continuous broadband spectrum, thus the radiation frequency spectrum of sequential circuit and information processing circuits will be the superposition of two kinds radiation frequency spectrum of inerratic narrowband and continuous broadband.

In computer and other digital information equipment, the clock pulse is the basis of interpretation of the digital information. It is regular and stable with standard waveform and strict phase relation. Clock pulse signal has the following functions:

(1) Clock pulse is the trigger signal of the digital circuit state change to command orderly transmission of the digital signal.
(2) Each circuit only sent and read the digital signals on the rising edge and falling edge of the clock pulse when data is transmitted.
(3) The correctness of information equipment data reading is determined by the waveform quality of the clock pulse.

For example, the pixel clock pulse is a very important clock signal in LCD panel. The frequency of pixel clock pulse is related to the working mode of the LCD panel, if the resolution of the LCD panel is higher, the frequency of the pixel clock pulse is higher. In one line, the number of pixel clock pulse is equal to the number of pixel of the LCD panel. For the LCD panel with 1024 × 768, there are 1024 pixels in each line, so the number of pixel clock pulse is also 1024 in the effective video range. Figure 3.12 shows corresponding relations between pixel clock pulse and RGB signal in LCD panel with 1024 × 768 resolution.

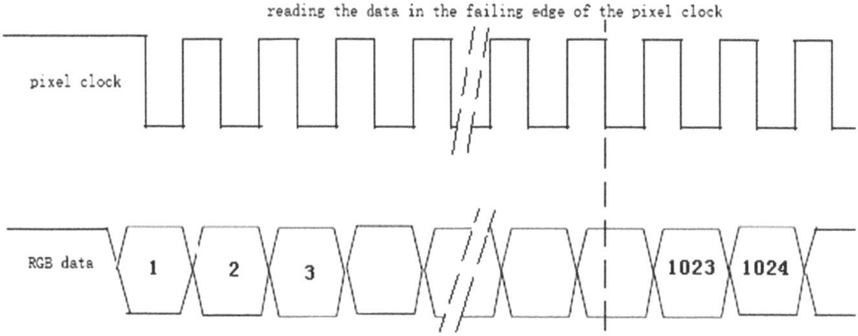

Fig. 3.12 Correspondences between the pixel clock pulse and the RGB signal

In fact, whether information can be reconstructed and reproduced depends on many factors, but the clock pulse is the most critical one. Clock pulse is necessary for interpreting the digital information, if it can be obtained in intercepted information, it is possible to reconstruct the data and reproduce the information, and vice versa. Thus, it can be concluded that clock pulse is one of the most important red signals.

3.2.3 Quality Assessment on Information Reproduction

Electromagnetic radiation from information equipment has a wide range of frequency band, and its electromagnetic waves may carry useful information, which is usually parasitic modulated into an electromagnetic wave and radiated out. For intercepting device, the obtained useful signal is leaked by information equipment which is often accompanied by various noise signals. How to ensure reproduction quality and obtain clear and distinguishable information is very important.

For achieving data reconstruction and information reproduction, five variables are introduced in chapter two, that is, I is for the total amount of original information processed in unit time by information equipment, C_{ko} denotes channel capacity of electromagnetic leakage information under the ideal condition, C_{k1} stands for channel capacity of electromagnetic leakage information under the noise condition, C_r is for the amount of information intercepted in unit time, and Q represents the quality of data reconstruction and information reproduction. The reconstruction and reproduction of video signal in the display are taken as an example to analyze the quality of image reproduction, which is shown as follows.

Quality of data reconstruction and information reproduction is

$$Q = \frac{C_r}{I} \tag{3.3}$$

the amount of original video information is

$$I = (\#ofcolors\,[\text{bit}]) \times (displayresolution\,[\text{pixel}]) \times (framerate\,[\text{fps}]) \tag{3.4}$$

In the formula, *#ofcolors* is the number of pixels, *displayresolution* is the resolution of display, and *framerate* is frame rate.

Intercepting channel capacity is

$$C_r = C_{k1} = k_1 B \log_2(1 + S/N) \tag{3.5}$$

The Intercepting channel capacity is equivalent to the channel capacity of electromagnetic leakage information under noisy condition, therefore

$$Q = \frac{k_1 B \log_2(1 + S/N)\,[\text{bps}]}{displayresolution\,[\text{pixel}] \times framerate\,[\text{fps}]}\,[\text{bit}] \tag{3.6}$$

The conclusion drawn by analysis is that there is a relationship between the video signal and channel capacity as follows:

(1) When $C_r \geq I$, it is possible to obtain information from the video signal and reproduce a clear image.
(2) When $C_r < I$, only part of the information can be obtained to reproduce a blurred image.
(3) The value of Q is a threshold value indicating whether the reproduced video image is clear and effective or not. Reproduction test implemented by a large number of intercepted video signal confirmed that when $Q > 0.8$, the reconstructed and reproduced image is clearly visible; when $Q > 0.5$, the image is basically visible; and in other cases, discernible image cannot be reconstructed and reproduced.

3.3 Techniques on Electromagnetic Information Reproduction

Data reconstruction and information reproduction from intercepted electromagnetic information require meeting certain conditions [17, 18]. First of all, whether it can clearly reconstruct the data and reproduce the information or not is closely related to the total amount of information intercepted. Meanwhile, as the clock pulse plays a very important role for sending and reading data in digital circuits, so only the clock pulse is found from intercepted information before the original information could be reproduced. Finally, the technique adopted for original information reconstruction also relies on understanding the working mechanism and data transmission format of specific equipment.

3.3.1 Video and Image Reproduction Techniques

To reproduce a clear video image, require intercepting enough amount of information first, and based on that, major synchronization control signal, such as horizontal, vertical, and data enable of the monitor, will be analyzed to obtain from intercepted information according to monitor display mode, then the video image could be tried to reproduce [19–24].

Analog RGB cable is commonly used for signal connection between monitor and host, whose data transmission format is shown in Fig. 3.13a. Each pixel is composed of a set of RGB colors, *t* in the figure represents the pixel clock, and R_{t-1}, G_{t-1}, B_{t-1}, R_t, G_t, B_t, R_{t+1}, G_{t+1}, B_{t+1} represents three adjacent pixels, respectively.

When data is being transmitted, the electromagnetic signal leaked by the cable is superimposed waveform of three RGB analog signals, which is shown in Fig. 3.13b, if the transmitted RGB pixel value in time $t-1$ is different from that in time t, there will be appeared peak value on the EM waveform, then leakage may be occurred.

$P_E(t)$, the difference of electromagnetic wave energy of the adjacent pixel, indicates the electromagnetic wave leakage energy at this time, and its estimation formula is

$$P_E(t) = c \cdot |p_t - p_{t-1}| \tag{3.7}$$

In the above equation: C is a constant and P_t is the electromagnetic wave energy of the pixel transmitted by pixel clock t.

The following Figs. 3.14, 3.15, 3.16, 3.17, 3.18, 3.19, and 3.20 is a detail reproduction case [25]. Figure 3.14 shows the original image and the leakage diagram estimated by Eq. (3.7) are shown in Fig. 3.15, Fig. 3.16 shows the actual effect of interception and reproduction. The actual reproduction diagram is basically consistent with the predicted leakage diagram, what we got from Figs. 3.15 and 3.16.

The major control signals of LED monitor are composed of horizontal synchronization, vertical synchronization, and data enable, which are all synchronized by pixel clock. The effect of these control signals is shown in Fig. 3.17.

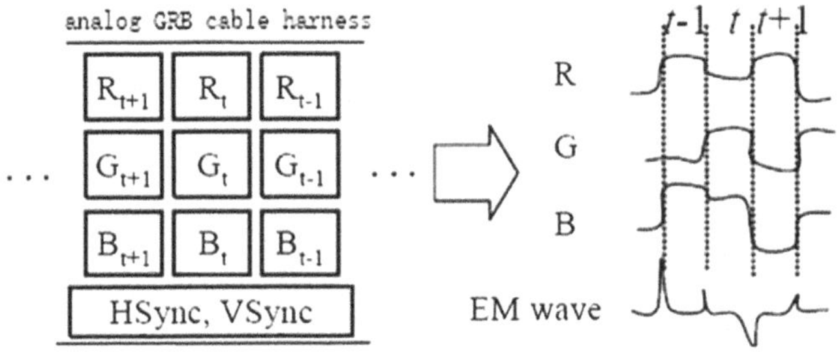

(a) Data transmission format of the analog RGB (b) Superimposed EM waveform of RGB analog signal

Fig. 3.13 RGB data construction and electromagnetic radiation wave during data transmission

Fig. 3.14 Original image

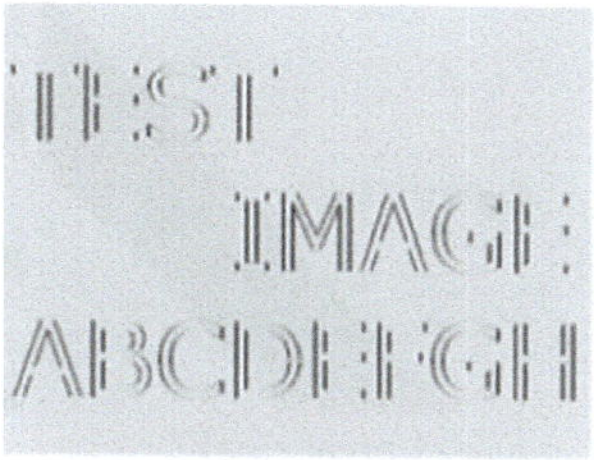

Fig. 3.15 Estimate leakage diagram

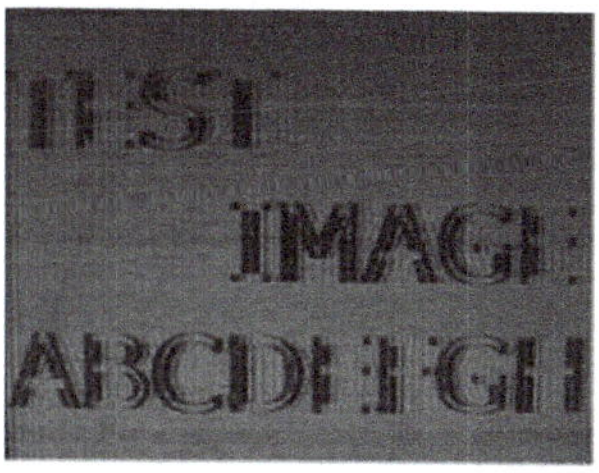

Fig. 3.16 Practical recurring figure

Table 3.1 shows the synchronous signal parameters listed in XGA display mode. When the monitor is in XGA mode, the synchronization control signal must be set up with the standards of Table 3.1. These clock synchronous signals are very critical in reconstruction and reproduction process.

Figure 3.18 shows the reconstruction and reproduction rendering in the case of horizontal synchronous signal changed while other parameters remain unchanged, among these, the horizontal synchronous signal modulation rate is the number of pixel clocks per line. It is shown that when the horizontal synchronous signal has

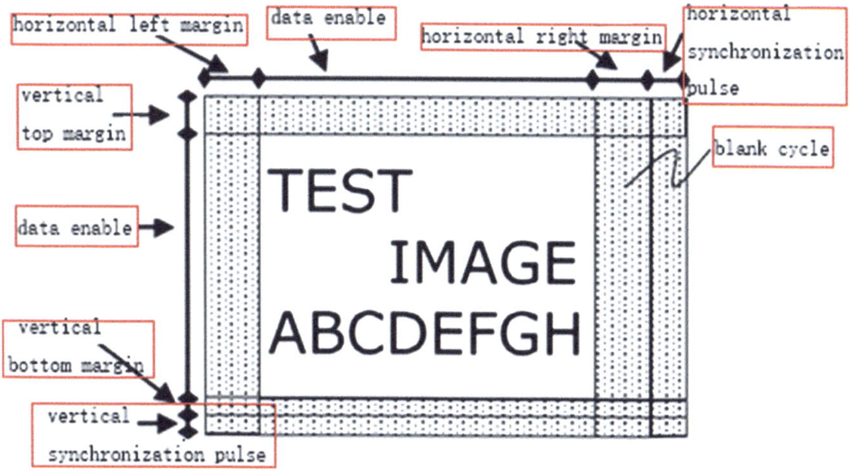

Fig. 3.17 The relationship between display image and synchronous signal

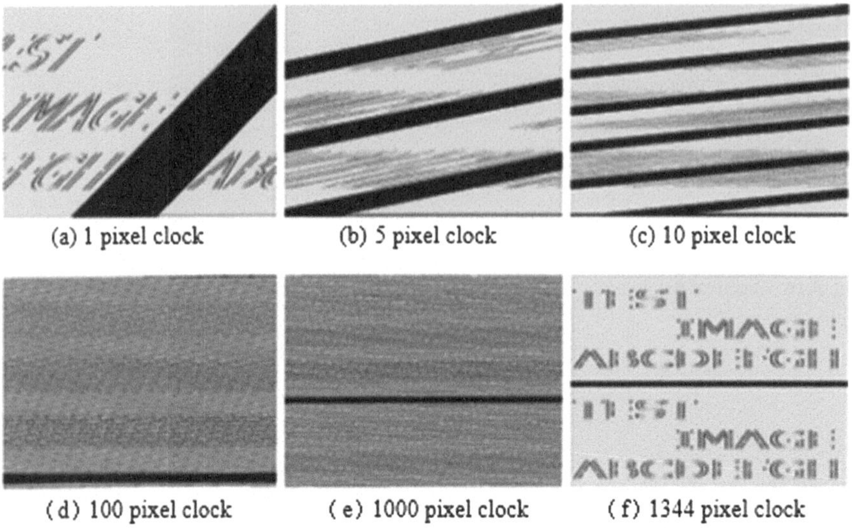

Fig. 3.18 Horizontal synchronous modulation of a reproduction image

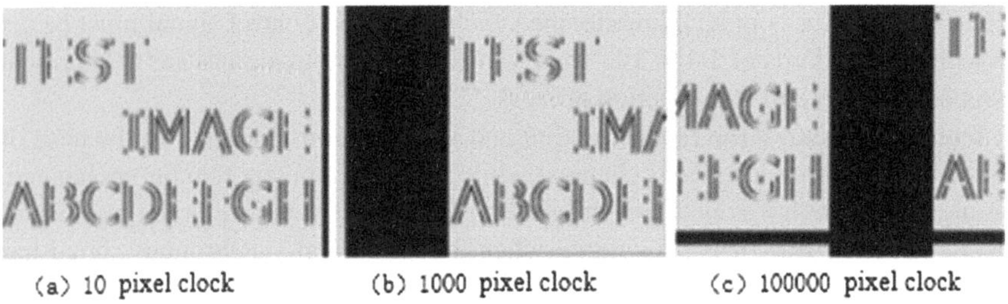

Fig. 3.19 Vertical synchronous modulation of a reproduction image

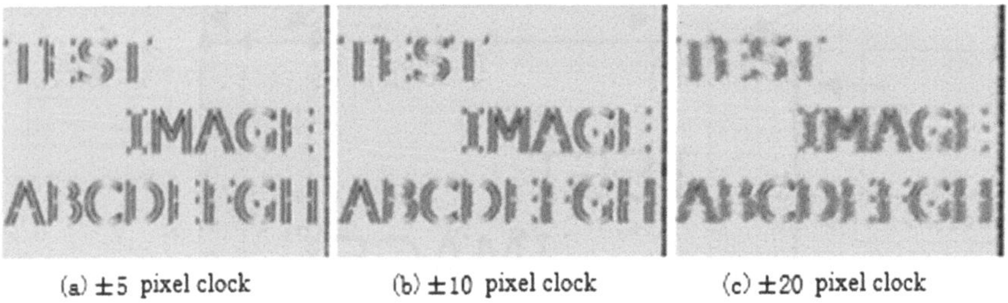

Fig. 3.20 Data enable synchronous modulation of a reproduction image

Table 3.1 Synchronous signal parameters of XGA display mode

Pixel clock/MHZ	Synchronization type	Parameters/Pixel clock			
65	Horizontal synchronization	Data enable period is 1024	Left margin is 24	Synchronization pulse is 136	Right margin is 160
	Vertical synchronization	Data enable period is 768	Top margin is 3	Synchronization pulse is 6	Bottom margin is 29

1344 pixel clocks, which is the sum of number 1024, 24, 136, and 160, as listed in Table 3.1, the best the reproduction effect is obtained.

Figure 3.19 shows the reconstruction and reproduction rendering in the case of vertical synchronous signal changed while other parameters remain unchanged, among these, the vertical synchronous signal modulation rate is the number of pixel clocks per screen. Combined with Table 3.1, it shows the best reproduction effect when vertical synchronous signal approaches to six.

Figure 3.20 shows the reconstruction and reproduction rendering in the case of data enable synchronous signal changed while other parameters remain unchanged, among these, data enable synchronous signal modulation rate is on the basis of 1024 pixel clocks for XGA display mode, which fluctuates randomly in the selected modulation range. As shown in the picture, the smaller the deviation between data enable synchronous signal and the given value of 1024 is, the clearer the reproduction effect is.

The result got from the above reproduction process is that the clock synchronous signal, which is based on pixel clock, plays a critical role in the reconstruction and reproduction process. Thus if the pixel clock, the most important clock pulse, can be found in intercepted data at the time of reproduction, it is possible to reproduce the original image clearly.

3.3.2 *Keyboard Data Reconstruction Technique*

The keyboard of information equipment is an important input device, which has a set of buttons. A large number of buttons in the keyboard are usually arranged in the form of rows and columns. Input and output ports of keyboard controlling chip are connected by rows and column lines. The keyboard is constantly scanned by the chip, such as successively changing the electrical level of each column and reading electrical level of the row to determine whether a button is pressed, in which process delaying to remove jitter is required, and then sends the scan code. Different types of keyboard transmission protocol may not be exactly the same, but they have the same working mechanism. To reproduce keyboard button information from intercepted information, the key is to obtain a clock pulse [26–30], and the mainly used techniques are described as follows:

1. Edge Transform Technology (ETT)

This technology uses clock-jumping edge to read data. The keyboard working mechanism and a large number of measured data statistically point out some certain characteristics of the keyboard leakage signal, that is, the signal falling edge has strong radiation, the clock falling edge and the data falling edge is adjacent but not overlapping, the clock falling edge has greater radiation intensity when data is in high electro level. These characteristics are used to interpret the intercepted waveform to deduce the keyboard scanning code. The keyboard reproduction examples of this technique are presented in Chap. 6.

2. Matrix Scanning Technology (MST)

This technique unscrambles intercepted data by using the keyboard row and column scanning mechanism. The keyboard is arranged in rows and columns with buttons, which results in a longer wire. When the keyboard is working, it is polling each column successively, which forms the interval pulse, and its time is greater than 3us. This pulse signal radiates the electromagnetic wave by antenna effects of keyboard column conduction wire, and the intercepted electromagnetic wave appears as an evenly spaced signal. When a button is pressed, the keyboard needs to determine which button and send the scanning code by calling the subroutine, this action causes the query time of next column delays. In that case, the delay area appears in the intercepted electromagnetic wave, and this delay signal can be used for reproduction analysis and is related to the pressed button. The following example shows the intercepted electromagnetic wave that used s loop antenna with a diameter of 1 m in the semi-anechoic chamber, which is 1 m away from the keyboard [30]. Figure 3.21 and Fig. 3.22 are the electromagnetic waves when button C and H is pressed, respectively.

3. Modulation Technology(MT)

The electromagnetic information model in Chap. 2 describes the signal processed by information equipment radiates electromagnetic wave through equivalent antenna in the work circuit. For the band-pass filtering characteristic of equivalent antenna,

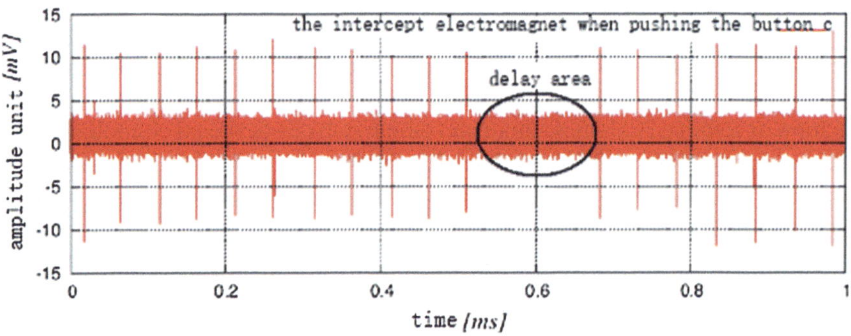

Fig. 3.21 Electromagnetic wave is intercepted when button C is pressed

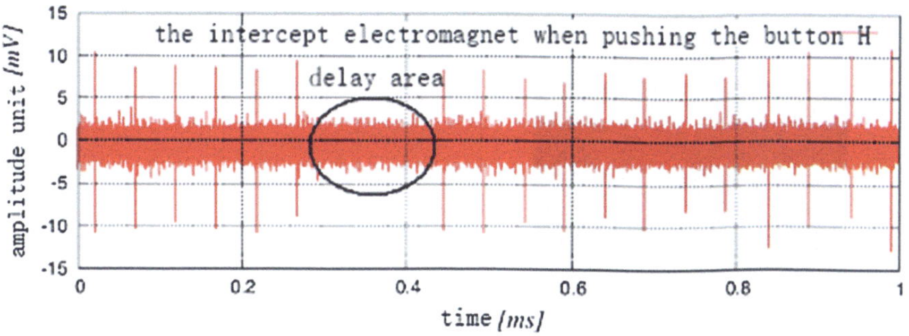

Fig. 3.22 Electromagnetic wave is intercepted when button H is pressed

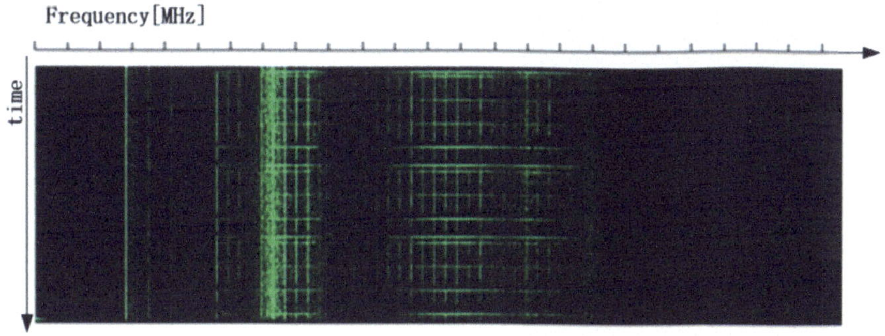

Fig. 3.23 Short-time Fourier transform of intercepted spectrum signals

radiation electromagnetic wave frequency is a part of harmonic of the information equipment's working clock, this principle can be used for keyboard button information reproduction. The following example of keyboard PS/2 illustrates the principle that achieves keyboard interception and reproduction by using the modulation technique.

Figure 3.23 shows the short-time Fourier transform of the intercepted spectrum signal when button E is pressed in the semi-anechoic chamber [30]. The upper limit of the transform frequency is 250 MHz, the Kaiser window is 40, 65,536 points, vertical line is the carrier frequency, and horizontal line is the keyboard scanning pulse.

Because the composition of intercepted electromagnetic waves is complex, such as the interference from clocks, nonlinear element, crosstalk, and ground, it is very difficult to distinguish the radiation source. However, it can be inferred the possible radiation source according to the working frequency of the component chip roughly. For example, the clock on the keyboard working chip is 4 MHz, and these electromagnetic waves are most likely leaked by the keyboard equivalent antenna when the harmonic frequency is multiples of 4 MHz. The frequency spectrogram centered on

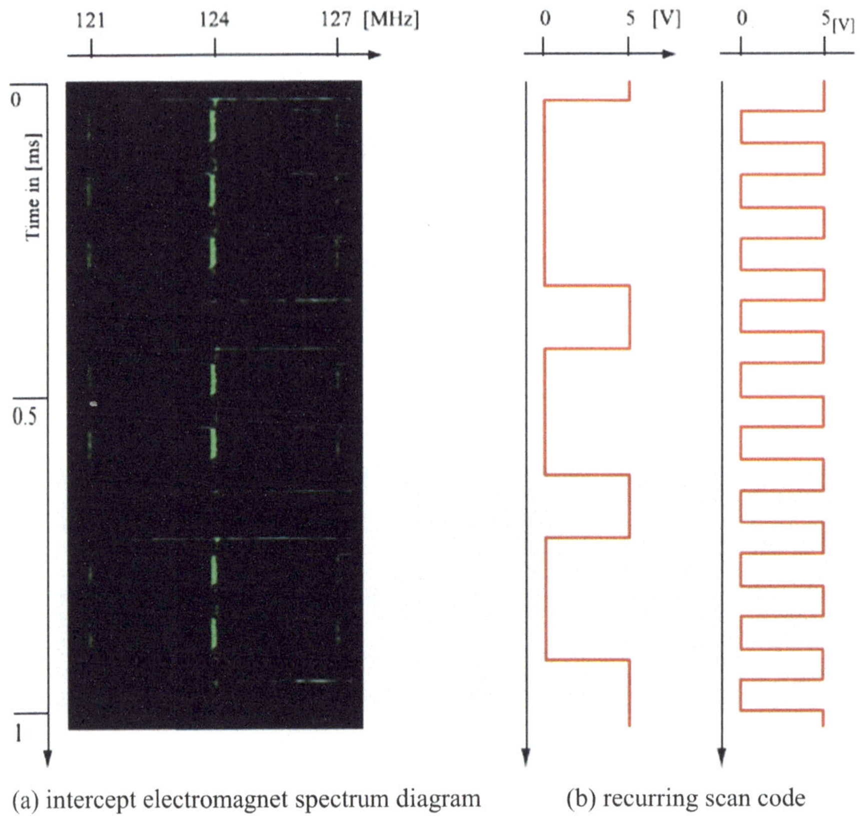

Fig. 3.24 Electromagnetic wave radiation and reproduction of button

124 MHz, which is frequency multiplication of 4 MHz, is intercepted from Fig. 3.23, which is shown in Fig. 3.24a.

To interpret harmonics from the relevance of data and clock signal perspective, the modulation signal of data on the clock can be clearly seen to describe the state of clock and data signal completely, thus the complete scanning code can be reproduced from harmonics, which is shown in Fig. 3.24b.

Each keyboard reproduction technique introduced above has its own characteristics, some of them have good reproduction effect, and some of them can be used in the long distance to intercept and reproduce. The following is the comparison of these techniques in the semi-anechoic chamber and office environment, which is shown in Figs. 3.25 and 3.26.

Effects comparison in other actual environments is similar with the office environment, if there are pipelines or cables, with their conduction and radiation action, the reproduction effect in other actual environments may be better than that of in office environment, even close to the effect of in semi-anechoic chamber.

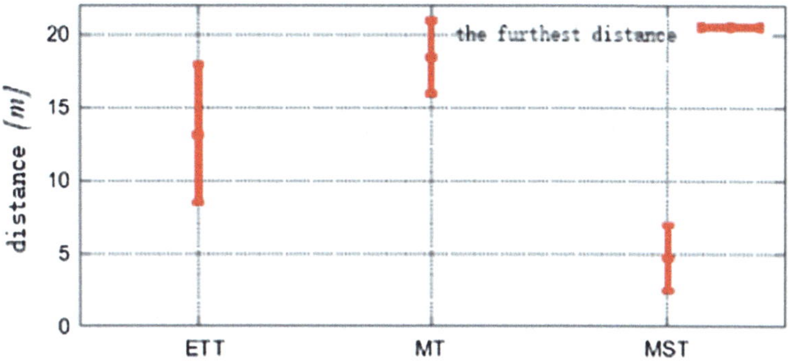

Fig. 3.25 The maximum distance of successful reproduction button by using various technologies in a semi-anechoic chamber environment

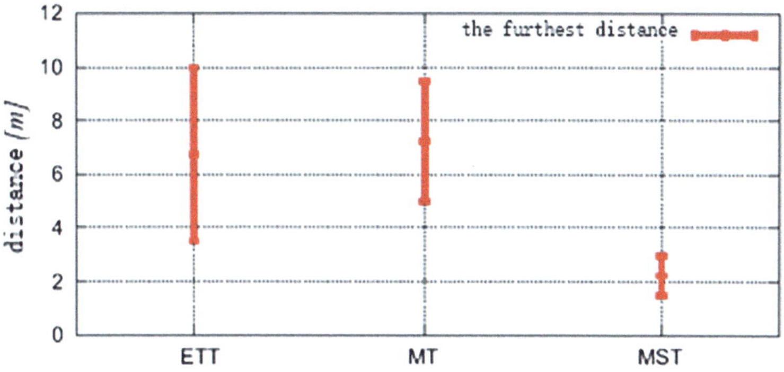

Fig. 3.26 The maximum distance of successful reproduction button by using various technologies in an office environment

3.3.3 *Laser Printer Information Reproduction Technique*

The working mechanism of the printer is different from monitor, printing job is one-time done, which is totally different from the monitor, and that is, the same image is regularly refreshed. It cannot use the image average technique to do reconstruction and reproduction, so the printer information reproduction is much more difficult. The printer electromagnetic information leakage mainly occurs on the data connecting line which connects the printer. Such as laser diodes wire of laser print.

Information reconstruction of the printer is the same as monitor, which needs to obtain the important clock synchronization signal in the intercepted signal. Such as commands for print start, print finish, and carriage return. Then, the information is reconstructed based on printer mechanism and original information is reproduced.

The following gives an example of information reconstruction for black–white laser printer [31]. The purpose of this example is to obtain the printer time-domain electromagnetic signal in the experiment environment of the semi-anechoic chamber, and the devices used are consist of a shielded loop antenna with 10 mm diameter, a preamplifier, and a digital oscilloscope, which are shown in Fig. 3.27. The related parameters are listed below,

Resolution of printer: 600 dpi,

Time of laser head to scan one line: 820 μs,

Bandwidth of the oscilloscope: 200 MHz,

Sampling rate: 2×10^9 per second,

Sampling point number: 100×10^6.

The distance between the antenna and the printer is 10 mm.

It prints three types of images on the A4 paper, which are black and white bars, characters, and color bars printed as a gray pattern, which are shown in Figs. 3.28, 3.29, and 3.30.

To reproduce printed image by an intercepted electromagnetic signal, it is necessary to obtain three synchronization signals to reconstruct and reproduce the original image, which includes a signal for the print start, print finish, and carriage return. The three images reproduced appear in Figs. 3.31, 3.32, and 3.33.

The displayed images shown in Figs. 3.31, 3.32, and 3.33 only reproduced 2.5% of original contents, it is because oscilloscope's memory is not enough for recording all sampling data. Decreased sampling frequency results in increased recording time. As Fig. 3.34 is shown, the obtained reproduced content reaches 10%, but the reproduced image is blurred.

The clarity of the reproduced image is determined by intercepting distance. From Figs. 3.35, 3.36, 3.37, 3.38, 3.39, and 3.40, which show different reproduced quality by different intercepting distances, respectively, where on the left are reproduced image intercepted from Fig. 3.28, and on the right are reproduced image intercepted from Fig. 3.29.

In summary, the three examples described in Sects. 3.3.1–3.3.3 prove the three conditions of interception and reproduction, which are amount of information intercepted, clock pulse, and equipment working mechanism. When the amount of information intercepted reaches the threshold value of clarity Q (Eq. 3.3) and equipment

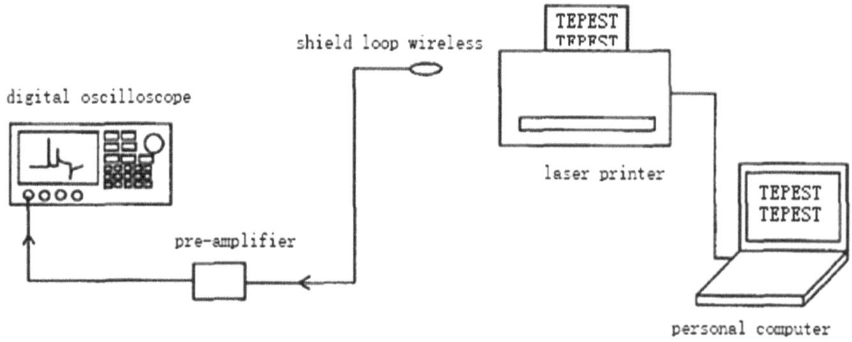

Fig. 3.27 The printer interception and reproduction device schematic diagram

Fig. 3.28 Print black and white bar image

Fig. 3.29 Print characters image

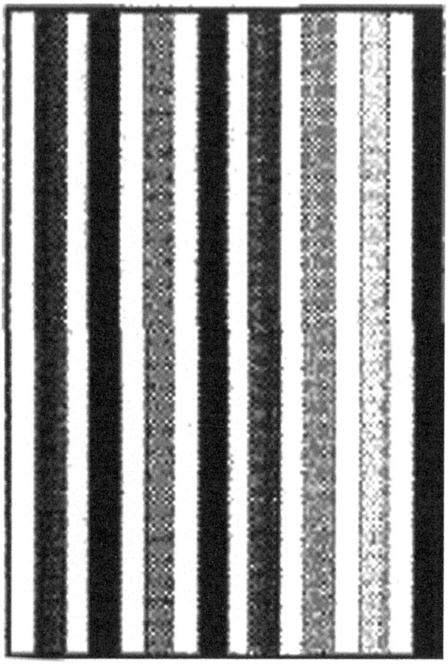

Fig. 3.30 Print color bar gray image

Fig. 3.31 Print reconstruct recurring image

Fig. 3.32 Print reconstruct recurring image

Fig. 3.33 Print reconstruct recurring image

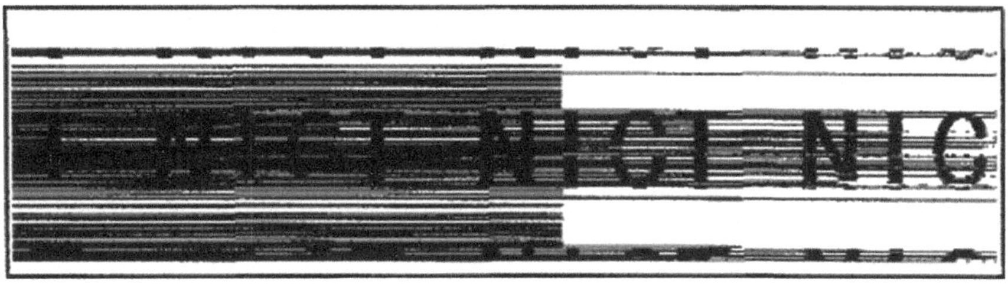

Fig. 3.34 Reduced sampling frequency recurring effect

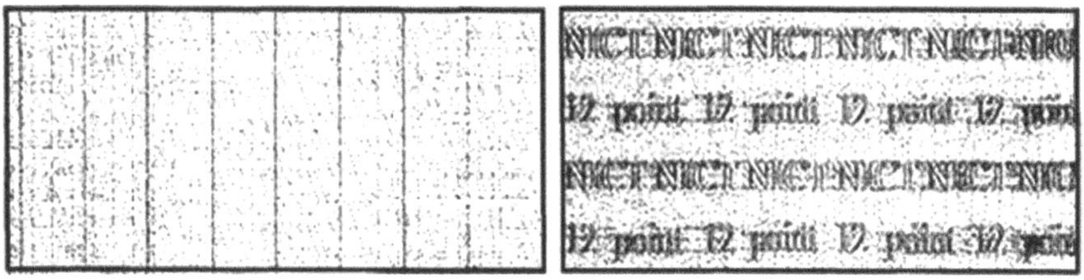

Fig. 3.35 0 cm intercept recurring diagram

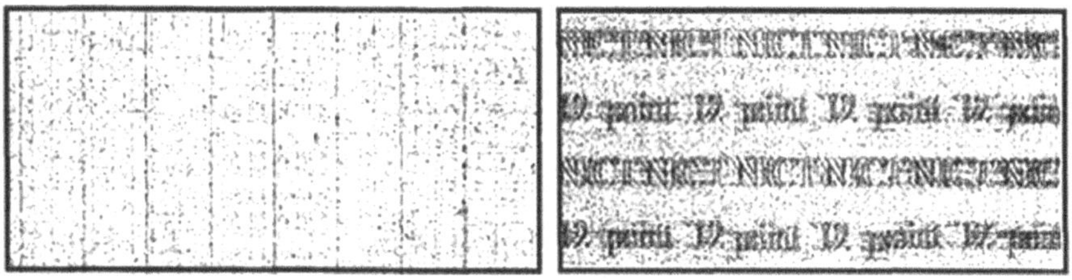

Fig. 3.36 10 cm intercept recurring diagram

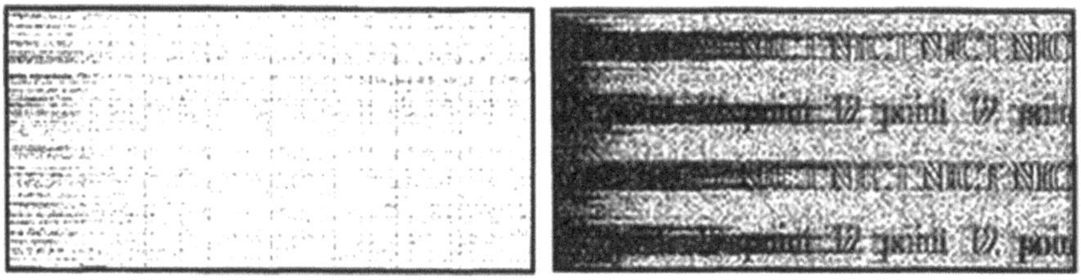

Fig. 3.37 30 cm intercept recurring diagram

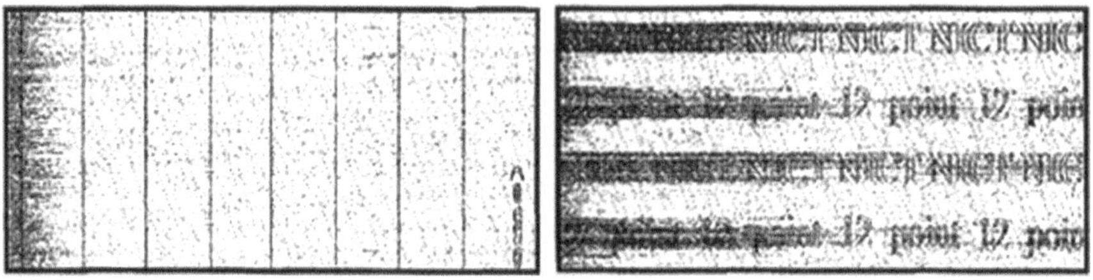

Fig. 3.38 50 cm intercept recurring diagram

Fig. 3.39 100 cm intercept recurring diagram

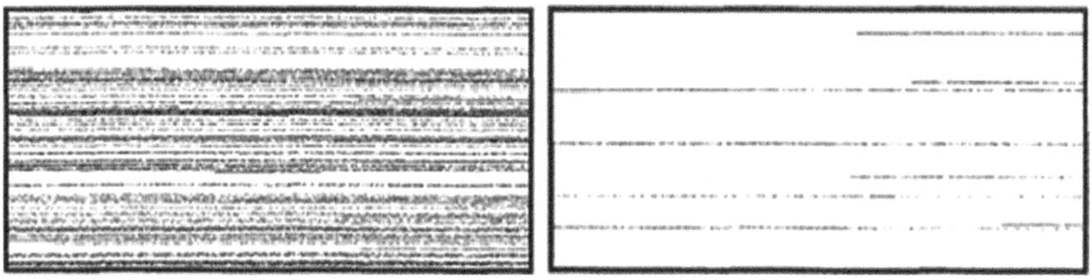

Fig. 3.40 200 cm intercept recurring diagram

working mechanism is known, determining clock pulse is critical. In the image reproduction case, clock pulse refers to horizontal synchronization, vertical synchronization, data enable, and pixel clock. Similarly, in keyboard reproduction case, the clock pulse is the synchronization signal in PS/2 protocol; in printer reproduction case, the clock pulse is the command of start, finish, and carriage return. Therefore, it is proved that the clock pulse is one of the critical red signals.

3.3.4 *Smart Card Secret Key Reconstruction Technique*

As mentioned in Sect. 2.1.3, electromagnetic signal and information are relevant. The change of processed information is represented by change of 0 and 1, leading to change of logic state for CMOS gate circuit, which causes the CMOS gate circuit's charging and discharging, thus electromagnetic radiation is generated. The Hamming distance of data reflects the change of information, the longer the Hamming distance, the bigger the information changes, and the stronger the electromagnetic radiates.

In two words, difference of corresponding bits is defined as the Hamming distance, that is,

$$\mathrm{H(x, y)} = \sum \mathrm{x[i]} \oplus \mathrm{y[i]} \tag{3.8}$$

In this formula: $i = 0, 1, \ldots, n - 1$, x and y are all encoding of bit n, $\oplus$ presents XOR. When x and y are binary codes, $H(x, y)$ is the difference number of bit between x and y.

The greater the Hamming distance of data, the bigger the current and power consumption, and the stronger the corresponding electromagnetic field, the more electromagnetic radiation is caused. Therefore, the correlation between the Hamming distance and amount of electromagnetic radiation can be established.

The letter E indicates the amount of electromagnetic radiation, which is

$$E = a * \mathrm{H(D, R)} + b \tag{3.9}$$

In the formula, Letter D presents the original data, R is result data, H(D, R) is Hamming distance between the original data and the result data, a and b is the correction coefficients.

When a = 1, b = 0, E is simplified as

$$E = \mathrm{H(D, R)} \tag{3.10}$$

Formulas 3.9 and 3.10 show the amount of electromagnetic radiation based on the Hamming distance. For the same register, before and after the command execution, the Hamming distance of two data determines amount of electromagnetic radiation E.

For data processed, if Hamming distance is not 0, which means that data is different before and after processing, resulting in a slight change in radiation energy. This change can be found by the differential statistical method, which can be used to infer the processed data is 0 or 1.

In electromagnetic information interception and reproduction, differential statistical methods are often used for power consumption and electromagnetic radiation, which is named DPA (Differential Power Consumption Analysis) and DEMA (Differential Electromagnetic Analysis) [32, 33]. Principles of the two methods are almost the same, that is, to collect a large amount of power consumption or electromagnetic radiation for different data processed, then analyze which by statistics method to find out data difference. The following is a smart card case that runs the DES encryption algorithm to introduce DEMA reproduction secret key method.

DES is a symmetric encryption algorithm, which uses iterative block cipher technique, it relies on a secret key to ensure data security. Smart card has the advantage of high speed, using DES algorithm for encryption and decryption. For encryption, the plaintext is grouped in 64 bits and the secret key is 64-bit, in which its valid secret key is 56-bit and parity check is 8-bit. The plaintext that needs encryption is divided into 64-bit data block, which is encrypted by 56-bit valid secret key. Each time encryption can do 16 rounds replacement and shift for 64-bit plaintext input, and output the totally different 64-bit ciphertext data, this process is shown in Fig. 3.41.

Input parameters of ith round iteration are denoted as L_{i-1} and R_{i-1}, which are, respectively, the 32-bit left half part and the 32-bit right half part, and then the ith iteration output is

$$L_i = R_{i-1}, R_i = L_{i-1} \oplus f(R_{i-1}, K_i) \tag{3.11}$$

The calculation of function f in the Formula 3.11 is shown in Fig. 3.42.

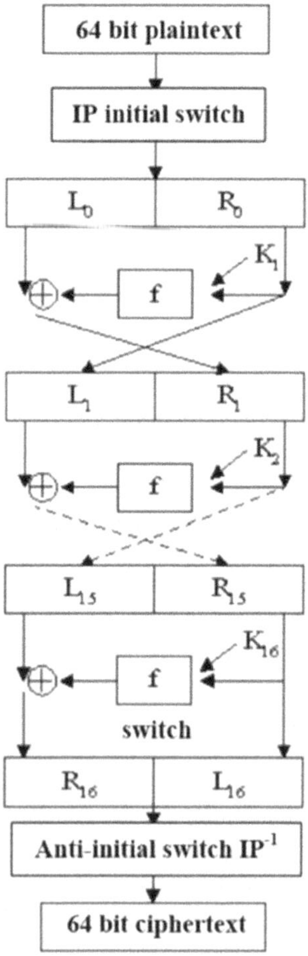

Fig. 3.41 DES encryption process

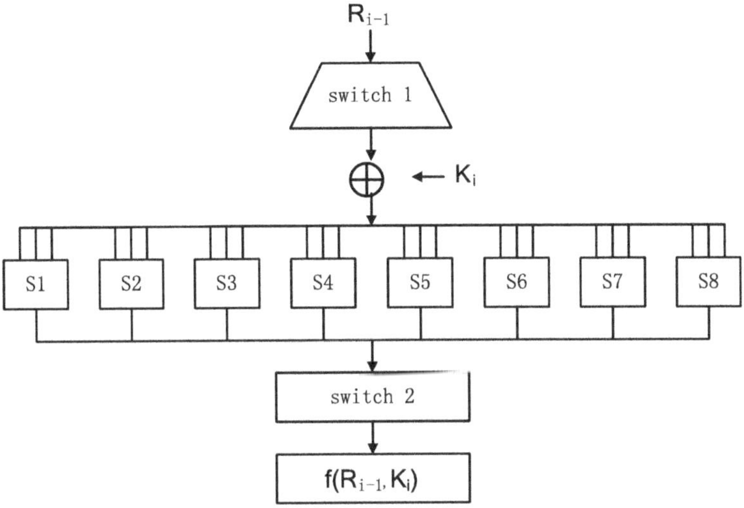

Fig. 3.42 Calculation process of $f(R_{i-1}, K_i)$

The output of ith iteration as the input of $(i + 1)$th iteration does the next time iteration. K_i is the 48-bit secret key that is confirmed by 56-bit secret key K; each time the iteration uses a different 48-bit secret key, and the production process of secret key K_i is shown in Fig. 3.43.

In Figs. 3.42 and 3.43, from the permutation 1 to permutation 4 is the permutation matrix required by the algorithm, and from the box S_1 to box S_8 is the eight permutation tables.

As the DES algorithm is a symmetric encryption algorithm, using DEMA to analyze and reproduce secret key has two methods:

(1) It records the plaintext and the corresponding electromagnetic radiation curve, by which the secret key is speculated.
(2) It records the ciphertext and the corresponding electromagnetic radiation curve, by which the secret key is speculated.

If it chooses the first method, the step of DEMA reproduction DES secret key is as follows:

(1) Data acquisition. The corresponding electromagnetic radiation curve will be generated when the determined plaintext is encrypted by using a same secret key, and then plaintext M_i and corresponding electromagnetic radiation curve S_i will be recorded as a group of data named "plaintext and electromagnetic radiation curve". Using the same secret key, the data set C will be formed by entering a large number of plaintext and group recording the "plaintext and electromagnetic radiation curve".

$$C = \{(M_i, S_i)|0 < i < N\} \tag{3.12}$$

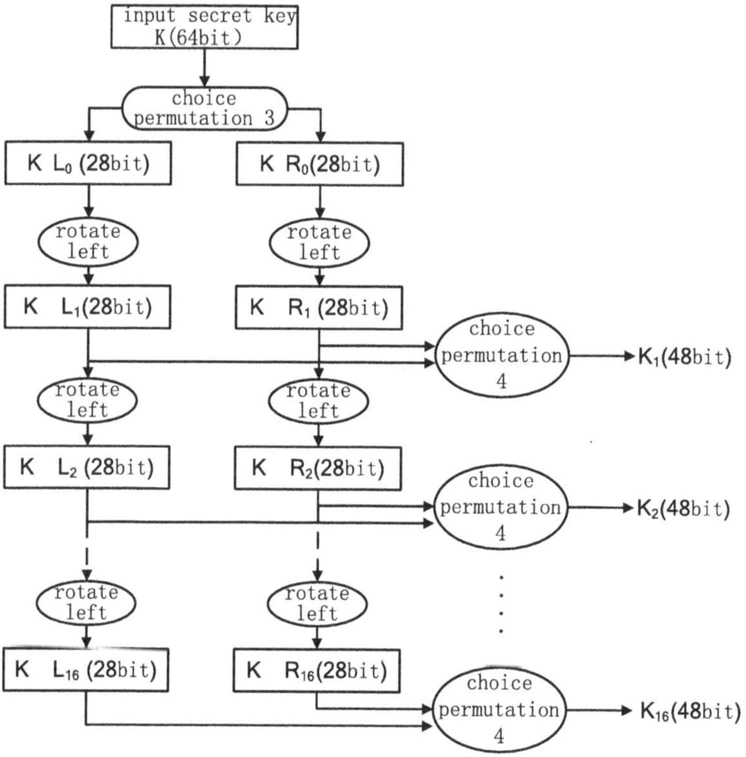

Fig. 3.43 The production process of secret key K_i

(2) Data analysis. It creates a classified determination function D of the plaintext and electromagnetic radiation curve, which is valued 0 or 1, and is related to plaintext M_i and secret key.

For example,

$$R_1 = L_0 \oplus f(R_0, K_1) \tag{3.13}$$

In above formula: L_0 and R_0 are left-half part and right-half part of plaintext M_i, respectively, and are all known. First round secret key K_i is 48-bit which is the attack target. For decreasing attacking difficulty, it can choose a part of the secret key K_i to attack, because

$$R_{1-1} = L_{0-1} \oplus f^*(R_{0-4}, K_{1-4}) \tag{3.14}$$

In above formula: K_{1-4} as a 6 bit sub secret key is a 6-bit inputting for fourth box S and chooses it as an attacking target.

Letting K_{1-4} as one estimated value, it calculates to get R_{1-1}, if $D = R_{1-1}$.

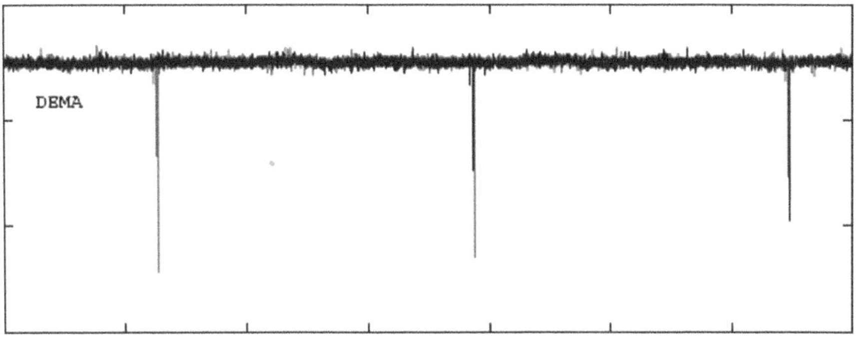

Fig. 3.44 Electromagnetic radiation difference curve

Using value D, it will divide the corresponding electromagnetic radiation curve of different plaintext inputting into two sets:

$$S_0 = \{S_i[j] | D = 0\} \tag{3.15}$$

$$S_1 = \{S_i[j] | D = 1\} \tag{3.16}$$

In above formulas: $|S_0| + |S_1| = N$, $S_i(j)$ is a jth sampling point value of ith electromagnetic radiation curve.
(3) The averaging value of the classification electromagnetic radiation curves is calculated

$$A_0[j] = \frac{1}{|S_0|} \sum_{S_i[j] \in S_0} S_i[j] \tag{3.17}$$

$$A_1[j] = \frac{1}{|S_1|} \sum_{S_i[j] \in S_1} S_i[j] \tag{3.18}$$

(4) It verifies the estimated sub-secret key by drawing the difference curve of electromagnetic radiation.
The curve R is obtained by calculating the difference between the two averaged electromagnetic radiation curves A_0 and A_1. If the guess of secret key is correct, then the difference value of curve R does not approach to 0 in corresponding time and shows obvious peak in Fig. 3.44 [33]. If the guess of secret key K is wrong, then the power consumption value of the curve R in the whole time is close to 0 without peak.

(5) Iterative process. The solution space of $K_{1\text{-}4}$ is $2^6 = 64$, and the correct $K_{1\text{-}4}$ can be obtained by only guessing for 63 times. And so on, the 48-bit sub secret key can be deciphered through differential analyzing the other 7 secret key segments. DEMA attacking for search space of sub secret key is decreased from 2^{48} to $8 \times 2^6 = 2^9$, there is no doubt that it significantly reduces the secret keyspace needs to search.

Chapter 4
Countermeasure Technology of Electromagnetic Information Leakage

Information security is inevitably threatened by electromagnetic leakage, which is becoming a more and more prominent issue with the wider application of information facilities in security and confidential departments, banks, financial field, and so on and the improvement of the intellectualization degree of crime. So, it is very important to master and effectively utilize the countermeasure for electromagnetic leakage. This chapter puts forward hierarchical protection strategy for electromagnetic information leakage with the aim of "anti-leakage" and "anti-interception" and "anti-reconstruction", and the new protection strategy is demonstrated in the form of specific cases on the basis of convention protection techniques.

4.1 Hierarchical Protection Strategy of Electromagnetic Information Leakage

According to the basic theory of electromagnetic information leakage proposed in Chap. 2, electromagnetic information leakage source, electromagnetic wave radiation and interception facilities correspond to signal source, signal channel, and signal sink, respectively. Therefore, electromagnetic information protection should comprehensively consider the three sources mentioned above: First, for the leakage source, electromagnetic leakage of information facilities should be blocked or suppressed as much as possible. Second, for electromagnetic information leaked, it is necessary to increase interception difficulty with the greatest degree. Finally, if the leaked electromagnetic wave were intercepted, make it difficult to reproduce is the right way to take. In a word, electromagnetic information protection should utilize the hierarchical protection strategy.

Hierarchical protection of electromagnetic information diagram is shown in Fig. 4.1, and Fig. 4.2 shows hierarchical protection measures.

"Anti-leakage" is the first level of protection strategy for electromagnetic information leakage, which mainly takes countermeasures for electromagnetic information

T. Liu and Y. Li, *Electromagnetic Information Leakage and Countermeasure Technique*,
https://doi.org/10.1007/978-981-10-4352-9_4

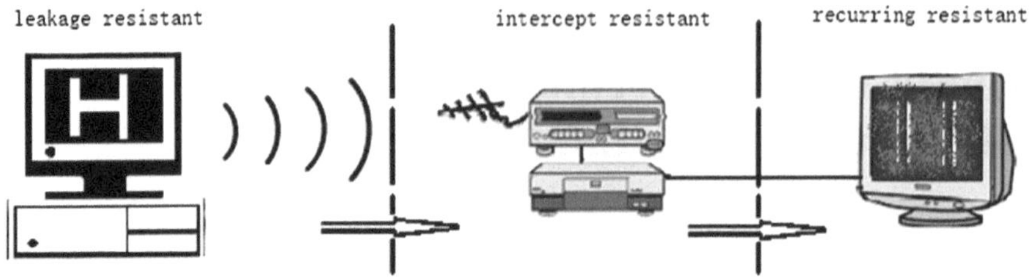

Fig. 4.1 Hierarchical protection of electromagnetic information diagram

Fig. 4.2 Hierarchical protection measures of electromagnetic information leakage

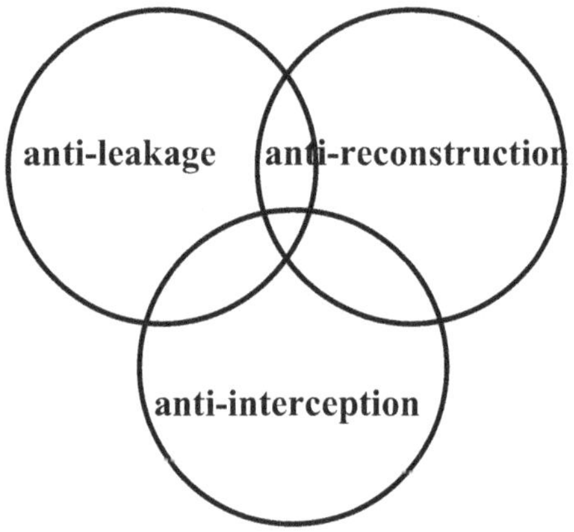

leakage equipment. By choice of equipment's low-radiation components and component model, combined with low-radiation reinforcement and shielding measures, in order to achieve low-radiation and low-leakage features of equipment, so as to block and suppress electromagnetic leakage. This level of protection is commonly used.

"Anti-interception" is the second level of protection strategy, which is applied in electromagnetic information radiation process. No matter what kinds of reinforcement measures are taken, more or less electromagnetic radiation will be emanated. Therefore, keep electromagnetic radiation signal minimum change or constant change will make interception of electromagnetic information much difficult.

"Anti-reconstruction" is the final level of protection strategy, which is protection countermeasures of intercepted electromagnetic information. The difficulty of data reconstruction and information reproduction is increased by signal variation (by methods of hardware and software) to prevent electromagnetic information reproduction.

For electromagnetic information leakage of information equipment, three levels of protection strategy can be adopted, which are correlative and complementary rather than completely independent of one another. In practice, for different equipment, applicable protection strategy can be determined by considering technique, cost, and effectiveness comprehensively.

4.2 Convention Protection Technology of Electromagnetic Information

As described in the above section, the hierarchical protection strategy of electromagnetic leakage is comprised of three levels, which are leakage, interception, and reproduction of electromagnetic information. The protection can be classified into hardware protection and software protection, other techniques include shielding, source suppression, and red and black separation. In order to achieve the purpose of electromagnetic information security, the hardware protection techniques are usually applying low-radiation facilities, shielding and strengthening, noise interference source, filtering, new materials and methods, and so on.

4.2.1 *Protection Type*

1. **Shielding**

For shielding protection, various hardware shielding measures of electromagnetic radiation in information facilities are applied, blocking the valuable electromagnetic signal or electromagnetic clutter interference signal to leak.

This type of protection is based on metal body to limit electromagnetic energy radiation by some measures like isolation, filter, electromagnetic absorption wave, shielding, and so on, suppressing leaked electromagnetic information through radiated emission and conducted emission [4]. The essence is to cut off the pathway of electromagnetic radiation to avoid electromagnetic information leakage, in which materials used including metal finger spring, electroconductive rubber, shielding wave band, ventilation waveguide, low-leakage coating, shielding grass, and so on [35–38]. The leakage source of electromagnetic information from information facilities is sealed to decrease the electromagnetic information leakage.

Shielding protection has been widely applied in many fields such as confidential environment and facilities. The shielded room is constructed in important locations, which including a confidential computer room, command and control center, meeting room, and so on. Electromagnetic shielding barrier is formed to prevent inside electromagnetic information leakage and prevent outside electromagnetic interference and destruction from the electromagnetic pulse. In addition, in the environment of maneuver and field operation, the electromagnetic shielding room, shelter, and shielding tent are applied, which can be rapidly installed and dis-installed to form mobile space of electromagnetic protection barrier, which can suppress electromagnetic leakage and emission from electronic equipment. Meanwhile, the immunity from interference and the capability of electronic countermeasure of information equipment can be strengthened.

The advantage of shielding protection is to carry out electromagnetic information protection according to existing information facilities and improve the original electromagnetic environment and promote the electromagnetic security level. However, the disadvantage of it is higher cost, larger weight, and bigger volume and inconvenience of deployment.

2. **Suppression Source**

Protection type of suppression source is an integrated optimization design on radiation source of the system included components, connector, and so on, which is a low-radiation design method adopted in the system design stage of information facilities in order to suppress or block the electromagnetic radiation of various radiation source of information system. In other words, it is an optimized design method for various electromagnetic radiation sources, such as system components, connectors, and various connection cables. The method includes low-radiation device selection, partial circuit design, and system-level optimization, to minimize potential sources of electromagnetic leakage during the system design phase.

Based on relative TEMPEST low-radiation standard, currently, the low-radiation products applied in protection type of suppression in the market include very large-scale integration chips, multilayer PC board, connector, low-noise components, low-radiation components and parts, shielding and absorbing wave devices and so on. The advantage of it is the radiation source have been suppressed at the original design stage of information facilities in order to decrease the workload of facilities protection operation at the later stage and reduce the protection cost and better system performance [39–41]. However, it is not applicable for all of the information facilities to apply the protection type of suppression, which should incorporate with other protection methods.

3. **Red and Black Signal Separation**

Protection of red and black signal separation is a kind of measure which the electromagnetic information leakage protection is only adopted for the red signal of information facilities.

According to Chap. 2, the electromagnetic signal can be classified into red signal and black signal, in which the red signal can be reproduced for useful information by interception, while the black signal cannot be. One of the important contents of TEMPEST is to distinguish and separate red and black signal in the system. The threaten of information security is mainly attributed to red signal, so identification of red signal is the key to determine whether or not compromising emanations during the test of electromagnetic information leakage occur.

Protection type of red and black separation only protect red signal leakage source by distinguishing between red and black signal according to the principle of red and black signal, which means hardware environment carrying or processing confidential information is protected. Black signal source generated the clutter signal, and useless signal is ignored. So, the advantages of the information facilities applied the protection type of red and black separation are lower cost, lighter weight, and higher efficiency; however, the electromagnetic compatibility or mutual interference issue is finitely solved.

4. **Noise Interference**

Noise interference is an active protection measure of information leakage, whose main target is "anti-reconstruction". The essence of which is mixture electromagnetic wave emitted from the jammer with electromagnetic wave signal radiated from the information facilities, so as to cover the original content and feature of leaked information. Even if the mixed signal were intercepted, the useful information can't be reproduced or reconstructed.

The noise interference source is demanded to relate with the electromagnetic feature of the protected signal, which the frequency spectrum, line field frequency, broadband frequency and so on should be synchronized as far as possible. Therefore, deployment of it is complex and the protection efficiency is low. Furthermore, because the new radiation electromagnetic signal is increased, the interference to the adjacent electronic facilities and influence on the health of the operator are caused.

4.2.2 Protection Technology

1. **Application of Low-Radiation Facilities**

Use of low-radiation facility (also called TEMPEST facility) is an important means to prevent radiation leakage, which reflects the strategy of "anti-leakage" in electromagnetic hierarchical protection. The United States is one of the earliest nations to research information leakage computer protection technology; its development of TEPEST technology also represents the world's advanced level. Shielding TEMPEST computers of U.S.A. dominate the whole business more than 10 years from NACSEM5100 standard published by NSA (National Security Agency) from 1970 to 1982. Red and black separation type of TEMPEST computer is another protection technology of electromagnetic information leakage following shielding TEMPEST computer after 1982. First, red and black separation applies system, box, and board level isolation to totally separate red signal from black signal. Second, the isolated red signal is processed by special measures to meet the extreme demands, while the black signal only meets EMC standards.

2. **Shielding Reinforcement Technology**

Shielding reinforcement can realize the target of electromagnetic protection by propagation suppression of electromagnetic radiation in space (cut off the channel of radiation), which reflects "anti-leakage" in electromagnetic hierarchical protection strategy. The essence is to shield the key components and circuit causing electromagnetic information radiation and the coupled electromagnetic field is attenuated by reflection and absorption, so as to achieve the purpose of "anti-leakage". The electromagnetic shielding uses metal material or various of metallization composite material to seal the electromagnetic interference source, which intensity of the outside electromagnetic field is lower than the allowable value or the electromagnetic sensitive circuits is sealed by metal shield material, which the internal electromagnetic intensity is lower than the allowable value. The following gives examples:

(1) Shielding box

Shielding box is very important for military facilities, which is focused on the aperture radiation issue. In fact, the box will have some holes and gaps, resulting in electromagnetic leakage, so that the shielding effectiveness is much lower than the theoretical calculation of the complete metal plate. The practice has proved that electromagnetic leakage is maximum when the dimension of hole and aperture equal to the integer multiple of half wavelength; in general, the aperture or hole length is $\lambda/100$–$\lambda/10$.

(2) Shielding reinforcement of display window

Leak-proof glasses or electromagnetic shielding window need to apply in order to prevent electromagnetic information leakage from the display. Electromagnetic shielding window is supported by the fine metal mesh between two layers of glasses or pressed into two-layer polypropylene and the metal film of vacuum deposition is processed on the optical substrate. The transmittance of the shielding window should be up to more than 32% in the condition of background light screen. At present, the frequency range of rapid flat window, flexible arc shielding window, and flexible planar shield window is from 9 kHz to 1.5 GHz; the shielding performance is up to 15–60 dB.

(3) Key circuit shielding

Local shielding can be applied in place of big radiation volume, such as shell ground of crystal oscillator. Little metal box can be used in CRT equipment, by which relating circuit of horizontal frequency, frame frequency and so on can be enclosed, respectively, to decrease radiation. Meanwhile, motherboard of the computer mainframe and various peripherals, power supply, and plugboard can be put in shielding boxes, respectively, which as a whole is then put into a larger shielding box.

3. **Noise Interference Technique**

Usually, interference source is placed near information equipment, so information radiation from information equipment is mixed with interference source; it is very difficult to reproduce the radiation information emanated from information facilities, which reflects "anti-reconstruction" in electromagnetic hierarchy protection strategy. In addition, electromagnetic radiation generated by jammer should not exceed the EMI standard. Currently, noise jammer is classified into two types:

(1) White noise jammer

White noise jammer is the early jammer product, strong noise is generated in a relatively wide frequency band to cover signal from electromagnetic information leakage in order to increase the difficulty of interception and reproduction. However, with the rapid development of digital signal processing technology, the noise signal can be filtered by digital signal processing, so the protection effectiveness of white noise jammer will be greatly decreased.

(2) Radiation relating jammer

Radiation relating jammer of information facilities can overcome the shortage of white noise jammer, which collects video radiation signal of the display to automatically radiate interference signal related to video radiation. This makes it difficult to reproduce the original electromagnetic information, thus this method not only improves protection effectiveness but also decreases the intensity of interference signal.

4.3 New Types of Protection for Electromagnetic Information Leakage

New types of protection for electromagnetic information leakage include pixel clock randomization, RGB color configuration, image noise adding, and TEMPEST font based on Fourier–Gaussian method.

4.3.1 Pixel Clock Randomization

When LCD displays an image, control signals like horizontal synchronization and vertical synchronization and data enable are needed, however, such synchronization signals are based on pixel clock. Usually, pixel clock in equipment is fixed, such as 25 MHz, 65 MHz, and so on. Once pixel clock is changed, many relating control signals are influenced too, this feature can be applied to information security protection.

The example of electromagnetic information protection based on pixel clock randomization is described as follows [25].

The data connection between the LCD display and mainframe is the leakage source. Before sending the image to display, randomizing pixel clock together with randomization of synchronization control signal of horizontal synchronization, vertical synchronization, and data enabling can prevent image reproduction, even if the leaked information is intercepted. The principle of pixel clock randomization is shown in Fig. 4.3.

The generation of random number is classified into a true random number and pseudorandom number according to authenticity, the former comes from true random source of physical world such as resistance in circuits or noise of device such as MOS, while the latter is generated from certain mathematical algorithm produced by computer, which the generation isn't a true random process compared with true random number.

The true random number is generated by direct amplification of circuit noise or frequency jitter of ring oscillator or unpredictability of hybrid circuit, while pseudorandom number is generated by the random process of computer simulation; com-

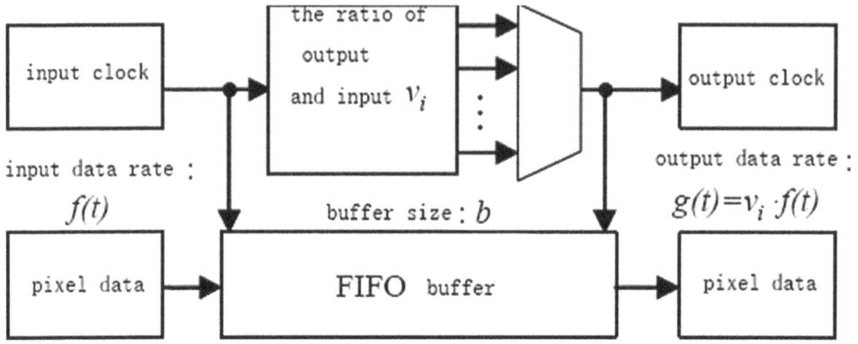

Fig. 4.3 Principle figure of pixel clock randomization

(a) original diagram (b) recurring diagram

Fig. 4.4 The comparison of unprotected original diagram and recurring diagram

monly used methods are LCG (Linear Congruent Generators), Tarsworthe displacement counting, Fibonacci delay generator, transition of pseudorandom sequence from chaos sequence, etc.

Pixel randomization needs to limit the value of random number generation. Value selection of output clock should satisfy two demands: one is to match the transmission rate of the original pixel, so that FIFO buffer is unlikely to overflow. The other is to display the image without causing image deformity.

Figure 4.4 and Fig. 4.5 show original and reproduced images without applying pixel clock randomization and protection measure, respectively. In contrast, pixel clock randomization technique can achieve better protection effectiveness.

4.3.2 RGB Color Configuration

Touch screen is widely applied in different information equipment, such as ATM, access control terminal, ticket selling machine, etc., often verification code or personal account password is required to enter, which demands protection of information security. However, the risk of electromagnetic information leakage exists during

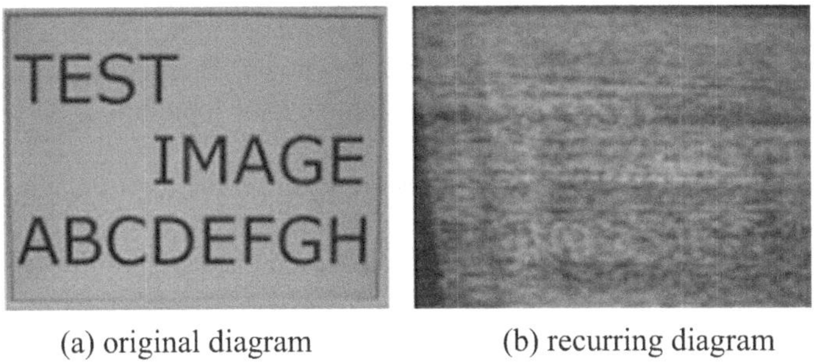

Fig. 4.5 The comparison of original diagram and recurring diagram after adapting pixel clock randomization

operation of touch screen. Once radiated electromagnetic wave from the touch screen is intercepted, the original information could be analyzed and image could be reproduced, so that security of electromagnetic information is threatened.

RGB color configuration is especially for information protection of touch screen and has better protection effectiveness [42].

The principle of different touch screens is generally the same, which adds a transparent touch panel to an ordinary liquid crystal screen. At present, touch screens can be divided into four types, which include resistance, capacitance, infrared, and surface acoustic wave. Typical resistive type of touch screen is composed of three parts, including resistivity conductor layer, isolation layer, and electrode, as shown in Fig. 4.6.

When it is working, the upper and lower conductor layers are equal to the resistance network, in which the voltage gradient will be formed when the voltage is added to the electrode of a certain layer. The voltage of contact point (as shown in character icon 1) can be measured at another layer without voltage in electrode if the upper and

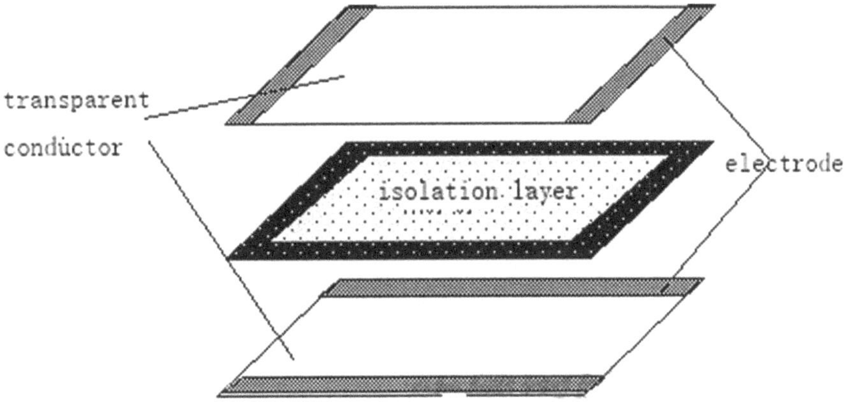

Fig. 4.6 Construction drawing of resistive type of touch screen

lower layers are touched in a certain point by the external force (the finger pressing), so that the coordinate of the contact point can be known. Then, from the interface to CPU, the input information can be confirmed; finally, the input information displays on the liquid crystal screen.

The radiation of touch screen essentially comes from the electromagnetic signal of RGB signal analog voltage variation during the raster scan process of liquid crystal screen rather than the digital graphic signal transmitted from graphic devices to liquid crystal screen.

The touch screen includes button area and display panel, in which button area shows an icon of input characters, and display panel shows input state.

Input validity is indicated by the change of color on the button icon, which is the commonly used model.

Image color of touch screen is determined by RGB (Red–Green–Blue) code, different colors corresponding to different RGB code, forming different combined voltage, variation of voltage generates electromagnetic radiation. The intensity of electromagnetic radiation is related to the voltage variation of neighbor pixels. Icon color changes when the data button is pressed, with the change of the combined voltage of RGB signal on corresponding adjacent pixels and radiation intensity. The bigger the intensity changes, the easier for listening and interception to succeed. Based on the above analysis, variation of RGB signal combination voltage is caused by a change of color by different RGB codes. In other words, different Hamming distances exist among various colors. The relationship between Hamming distance and electromagnetic radiation is illustrated in the following according to Fig. 4.7 [42].

1. **Original color scheme**

The color scheme is shown in Fig. 4.7; the color change of adjacent pixels on the touch screen from left to right is as follows:

Unpressed:
black (background) → green (icon) → white (character) → green (icon)

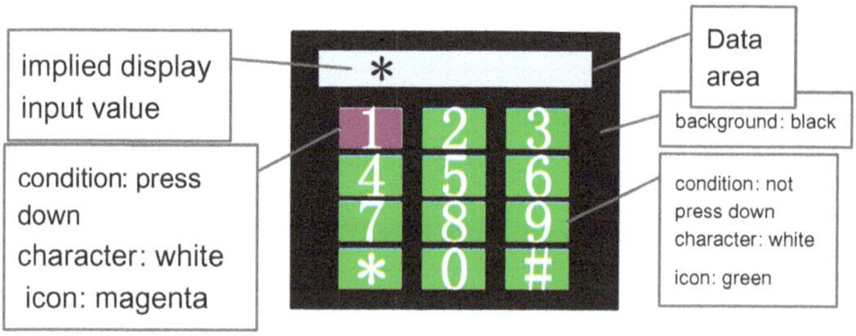

Fig. 4.7 Conventional color of touch screen

→ black (background).
Pressed:
Black (background) → magenta (icon) → white (character) → magenta (icon) → black (background)
in which RGB code of different colors is, respectively:
black [00H:00H:00H]; green [00H:FFH:00H];
white [FFH:FFH:FFH]; magenta [FFH:00H:FFH].
RGB Hamming distance caused by color variation is shown in Table 4.1.

2. **Leakage reason analysis**

In the anechoic chamber, the electromagnetic wave is detected from 1 m distance to touch screen by the antenna, as shown in Fig. 4.8; button icon on the touch screen is pressed to input password, the electromagnetic radiation signal from the touch screen is received by the intercepting devices; through data reconstruction by PC, the pressed button icon can be distinguished after information reproduction. The intercepted electromagnetic wave is shown in Fig. 4.9. From Fig. 4.9, before and after the button icon is pressed, the reproduced image is discernible: when button "1" is pressed, the vertical border of icon "1" becomes thicker, while the vertical line of character "1" becomes thinner, which can be used to predict that character "1" is pressed.

Before data input, Hamming distance between background color (black) and icon color (green) is eight, while when the button is pressed, the Hamming distance changes to sixteen between the background color (black) and icon color (magenta). So, the electromagnetic radiation intensity of the touch screen becomes stronger when the icon is pressed, the vertical border obtained from interception and reproduction is thicker than before the icon is pressed. Similarly, Hamming distance is sixteen between icon color (white) and character color (green) before data input, however, the Hamming distance is eight between the icon color (black) and character color (magenta) when the button is pressed. So, the electromagnetic radiation intensity of the touch screen is weaker than before the icon is pressed, the character vertical line obtained from interception and reproduction is thinner than before the icon is pressed. The test data verifies the above analysis results: the larger Hamming distance is, the stronger the electromagnetic radiation intensity is. It should be noted that the vertical line is obvious in the reproduced image and without electromagnetic leakage in the horizontal line. For the same reason, RGB combined voltage is stable

Table 4.1 Hamming distance of icon and character for color scheme in Fig. 4.7

Icon status	Icon color	Hamming distance	Edge color change	Hamming distance
Unpressed	Black → green → black	8	Green → white → green	16
Pressed	Black → magenta → black	16	Magenta → white → magenta	8

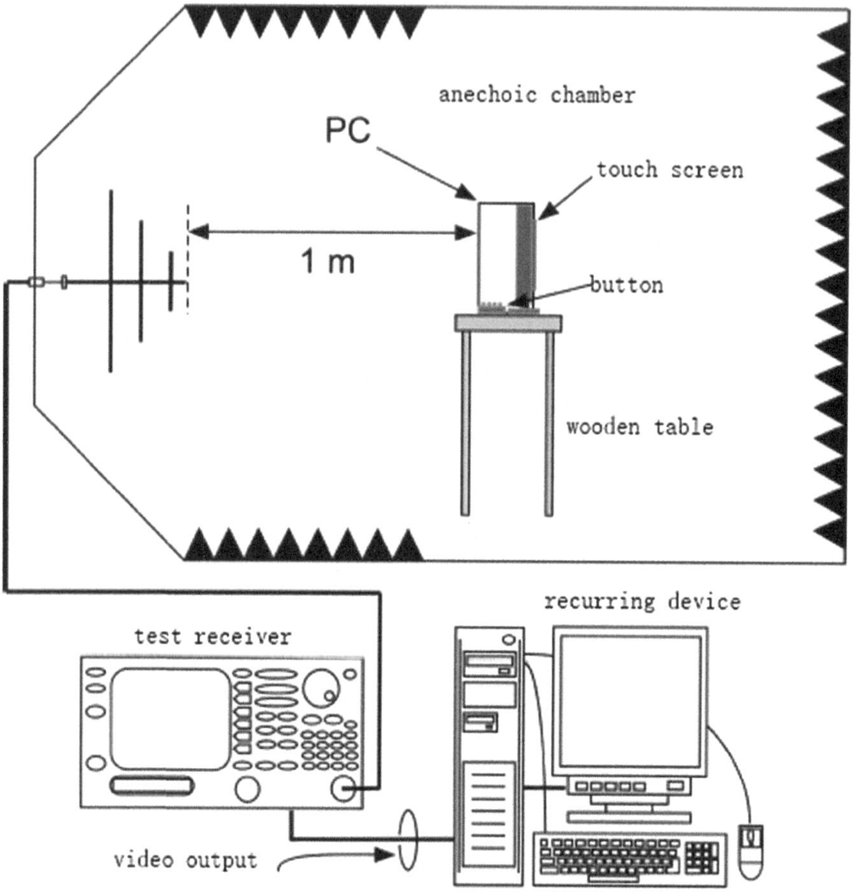

Fig. 4.8 Receive touch screen button electromagnet radiation system construct

in the horizontal direction, and Hamming distance of adjacent pixels is 0 without electromagnetic radiation, so no electromagnetic information leakage is caused.

3. **Protection principle and security color scheme**

According to the hierarchical protection strategy of electromagnetic information leakage from Sect. 4.1, it is improved that minimum variation of the electromagnetic radiation signal can avoid electromagnetic information leakage.

Protection principle for touch screen is by adjustment of configuration scheme of icon color, keeping Hamming distance of button, and digital border color variation constant before and after button is pressed. Though color changes before and after button is pressed, electromagnetic wave radiation keeps constant, so that the state cannot be distinguished before and after icon is pressed, so as to achieve protection of electromagnetic information leakage.

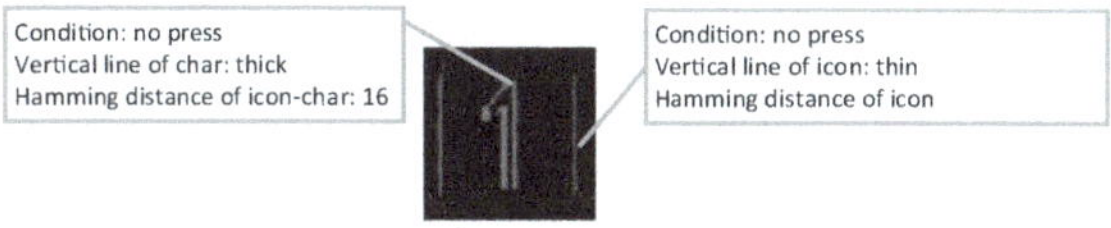

(a) icon before press down

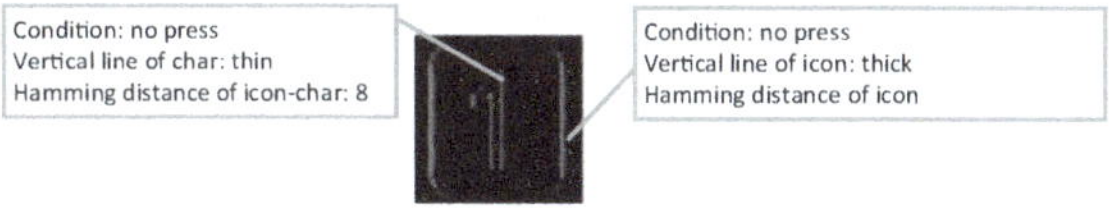

(b) icon after press down

Fig. 4.9 Comparison of reconstruction image before and after icon is pressed

The implementations and corresponding protection effects are explained by the following three kinds of color schemes.

Security color scheme A: the color pressed icon is adjusted to blue, RGB code is [0, 0, 255], as shown in Fig. 4.10, the color variation is shown as follows before and after the button is pressed:

Without pressing: black (background) → white (character) → green (icon) → black (background).

Pressed: black (background) → blue (icon) → white (character) → blue (icon) → black (background).

Here, RGB codes of different colors are shown, respectively, in the following: black: [00H:00H:00H]; green [00H:FFH:00H]; white [FFH:FFH:FFH]; blue [00H:00H:FFH]. RGB Hamming distance caused by color variation is shown in Table 4.2.

Fig. 4.10 Security color A of touch screen

Table 4.2 Security color scheme A icon and data Hamming distance

Icon status	Icon color	Hamming distance	Edge color change	Hamming distance
Unpressed	Black → green → black	8	Green → white → green	16
Pressed	Black → blue → black	8	Blue → white → blue	16

Security color schemes B: the color pressed icon is adjusted to the cyan, RGB code is [0, 255, 255], as the Fig. 4.11, the color variation relation is shown as follows before and after pressing:

Unpressed:
black (background) → cyan (character) → white (character) → black (background).
Pressed:
black (background) → yellow (icon) → white (character) → yellow (icon) → black (background).
in which RGB code of different colors is, respectively, shown the following: black, [00H:00H:00H]; yan, [00H:FFH:00H]; white, [FFH:FFH:FFH]; yellow, [FFH:FFH:FFH] RGB Hamming distance caused by color variation is shown in Table 4.3.

Security color schemes C: the color pressed icon is adjusted to the deep blue, RGB code is [0, 139, 139], as the Fig. 4.12, the color variation relation is shown as follows before and after pressing:

Without pressing: black (background) → green (icon) → white (character) → green (icon) → black (background).

Pressed: black (background) → cyan (icon) → white (character) → deep blue (icon) → black (background).

Fig. 4.11 Security color B of touch screen

Table 4.3 Security color scheme B icon and data Hamming distance

Icon status	Icon color	Hamming distance	Edge color change	Hamming distance
Unpressed	Black → cyan → black	16	Cyan → white → cyan	8
Pressed	Black → yellow → black	16	Yellow → white → yellow	8

Fig. 4.12 Security color C of touch screen

Table 4.4 Security color scheme C icon and data Hamming distance

Icon status	Icon color	Hamming distance	Edge color change	Hamming distance
Unpressed	Black → green → black	8	Green → white → green	8
Pressed	Black → cyan → black	8	Cyan → white → cyan	8

in which RGB code of different colors are shown, respectively, in the following: black, [00H:00H:00H]; Green, [00H:FFH:00H]; white, [FFH:FFH:FFH]; cyan, [00H:8BH:8BH]. RGB Hamming distance caused by color variation is shown in Table 4.4.

4. **Further analysis of color scheme**

Electromagnetic information leakage is generated in the original color scheme, which is shown in Table 4.1, electromagnetic radiation variation of the color scheme shown in Tables 4.2, 4.3 and 4.4 can keep constant with that of before and after when data is input in order to prevent from electromagnetic information leakage. However, in fact, the generated electromagnetic information radiation is different, which corresponds to different interception difficulties. It is known from the above analysis that the jumping of "0 → 1" or "1 → 0" is the main reason to generate the electromagnetic radiation peak and also the main component of the intercepted radiated electromagnetic wave. At the same time, it is also the key for data reconstruction and information reproduction. The frequent flipping between 0 and 1 will cause insecurity factors, so

Table 4.5 Comparison between 0 and 1 jumping of three kinds of security color schemes

Security color schemes	RGB code	0–1 flip
Schemes A Unpressed: black → green → white Pressed: black → blue → white	Black: 0000 0000 0000 0000 0000 0000 Green: 0000 0000 1111 1111 0000 0000 Blue: 0000 0000 0000 0000 1111 1111 White: 1111 1111 1111 1111 1111 1111	Icon unpressed 3 flips Icon pressed 1 flip Flip overlap
Schemes B: Unpressed: black → cyan → white Pressed: black → yellow → white	Black: 0000 0000 0000 0000 0000 0000 Cyan: 0000 0000 1111 1111 1111 1111 Yellow: 1111 1111 1111 1111 0000 0000 White: 1111 1111 1111 1111 1111 1111	Icon unpressed 1 flip Icon pressed 1 flip Flip overlap
Schemes C: Unpressed: black → green → white Pressed: black → cyan → white	Black: 0000 0000 0000 0000 0000 0000 Green: 0000 0000 1111 1111 0000 0000 Cyan: 0000 0000 1000 1011 1000 1011 White: 1111 1111 1111 1111 1111 1111	Icon unpressed 3 flips Icon pressed 9 flips Flip no-overlap

the scheme of less jumping is chosen between 0 and 1 in security color scheme of electromagnetic radiation protection, as shown in Table 4.5.

The more jumping times between 0 and 1, the more notable for the formed electromagnetic feature; the more serious electromagnetic leakage will be caused. It is known from Table 4.5 that due to minimum jumping times, scheme B is the most optimized one to prevent electromagnetic information leakage compared with the other two schemes.

Furthermore, electromagnetic radiation intensities are compared between security color schemes and conventional color scheme, which is shown in Fig. 4.13.

4.3.3 Image Noise Adding

Image noise adding is a kind of electromagnetic protection technique, which adds noise into the original image in order to randomize electromagnetic leakage signal and decrease SNR (Signal-to-Noise Ratio) of leakage signal.

It utilizes color addition feature without decreasing image quality after random noise is added, by which a random number is added into each pixel of the original

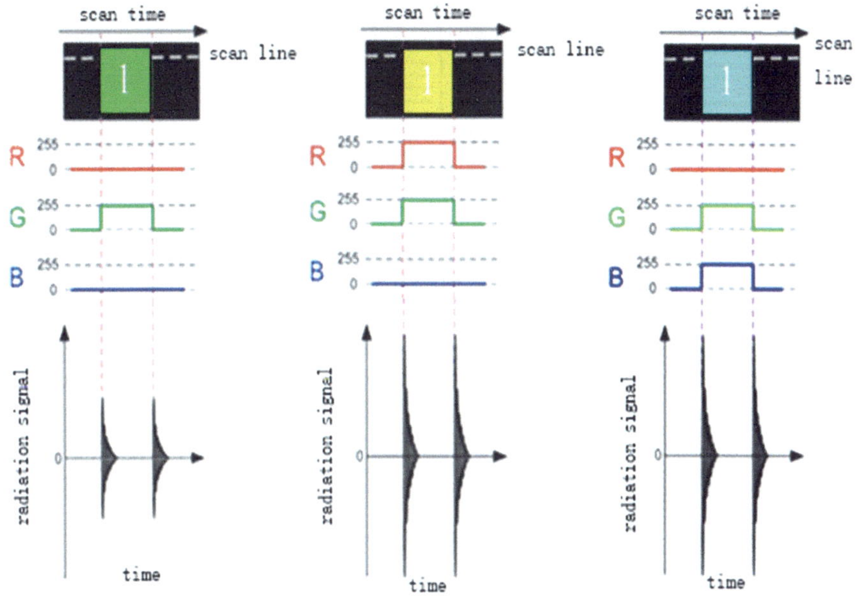

Fig. 4.13 Comparison of electromagnetic radiation intensity between security color schemes and conventional color scheme

image to generate the first sub-image, and the second sub-image is generated by subtraction of the same random number. The average pixel value of two sub-images is equal to that of the original image. The first sub-image and the second sub-image is rapidly shifted in the screen, human eyes observe the average display effect of correspondence pixel with both sub-image, so it seems that the observer still sees the original image [43].

Image noise adding embodies "resistance of reproduction" in electromagnetic protection strategy.

By analyzing the operation process of analog and digital display facilities, it is concluded that electromagnetic information leakage is generated by current variation, which caused by Hamming distance of adjacent pixel values. It randomizes electromagnetic leakage information by randomization of the pixel value for the purpose of electromagnetic protection.

For display facility with w by h resolution ratio, p_i expresses the pixel value of ith raster of original image scanned from the initial point (0, 0). u_i and v_i express pixel value corresponding to the first sub-image and the second sub-image, respectively. The arithmetic of Image noise adding is described as follows:

(1) Three images with size of $w \times h$ are produced, in which $w \times h$ of 8-bit random digit r_i is uniformly distributed in aggregates. (Only one aggregate required in gay scale image).
(2) Read input value p_i.

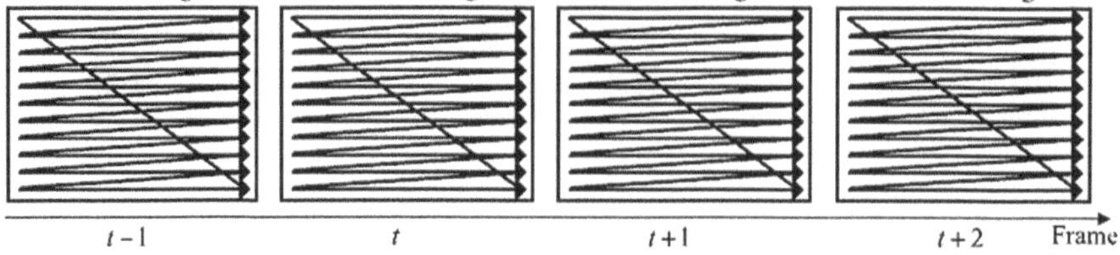

Fig. 4.14 Original image is displayed by two sub-images

(3) Each pixel value in three RGB color channels of the original image correspond to the random number in three aggregates produced in step (1), and implements loop execution for the following process ① and ②.

$$u_i = p_i + r_i$$
$$v_i = p_i - r_i$$

(4) u and v are taken turn to show by the speed of 1/60 s or faster and repetitive operation to show *p*, which is shown in Fig. 4.14.

Next example is to introduce the effectiveness of electromagnetic information protection by Image noise adding technique [25].

1. **Testing device and environment**

The testing applies FPGA device with IO interface to implement image noise adding. The input terminal of FPGA device is connected directly with the video output of the computer, and the output terminal is connected with CRT or LCD displays. Such a device is capable of real-time processing with image with XGA resolution ratio. The system and testing environment are shown in Fig. 4.15, in which the testing environment is comprised of two subsystems, including target system and monitor system; they are connected only with electromagnetic wave radiation. The electromagnetic radiation is captured by broadband receiver searching in full range (10 MHz–1 GHz). The power supply of the target system and monitor system should be isolated with each other because the current information will be propagated between the power supplies.

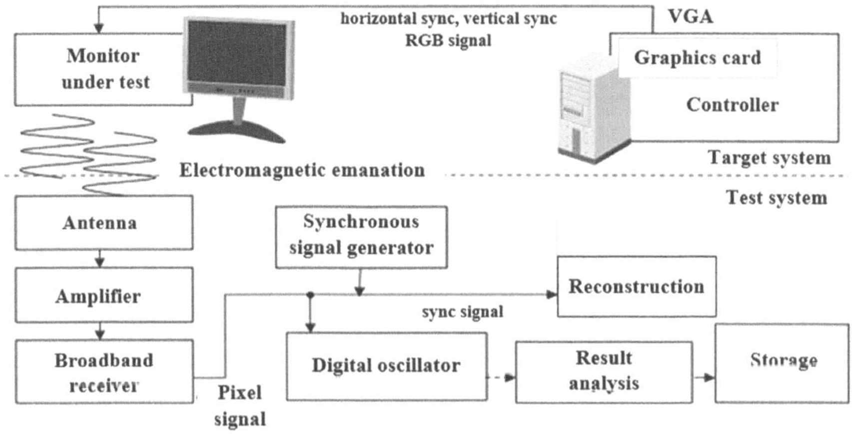

Fig. 4.15 Test environment diagram of electromagnetic leakage protection based on image noise adding

Table 4.6 Sync signal frequency of target devices

Target device	Hsync (kHz)	Vsync (Hz)
A (CRT)	48.47327212	75.0962
B (LCD)	48.38397523	60.0987
C (LCD)	48.53026104	60.0248

2. **Pretesting**

When horizontal and vertical sync signals generated by monitor are not coincided with the target device, the displayed image will roll in one or two directions. Keeping image stable, sync signal should be matched to 6–7 bits accuracy; the frequency of the target device can be obtained by pretesting.

The pretesting target is computer and display connected with analog RGB cable. One set of CRT display and two sets of LCD displays are tested, the results are listed in Table 4.6. The horizontal and vertical sync signal frequencies of the three devices are recorded in the table.

It is found that the electromagnetic information leakage of the target device is distributed among the frequency range of 10 MHz–1 GHz, which is shown in Fig. 4.16. According to the figure, the frequency produced the clearest leakage image is in 65, 134, 250, 327, 459 and 584 MHz; these frequencies can be regarded as the harmonic frequency of sync clock frequency.

3. **Testing results**

Two sub-images are produced by image noise adding, as shown in Fig. 4.17. The enlarged display of alphabet "s" is shown in Fig. 4.18, "s" of the original image is shown in Fig. 4.18a, and "s" of two sub-image produced from the original image

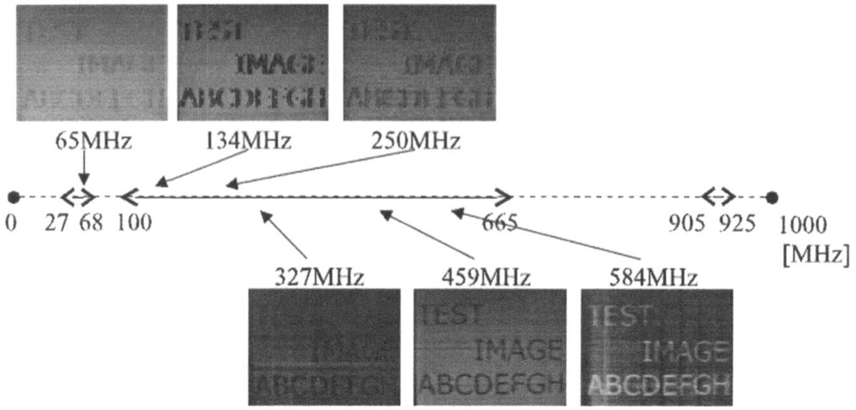

Fig. 4.16 Leakage image and leakage frequency

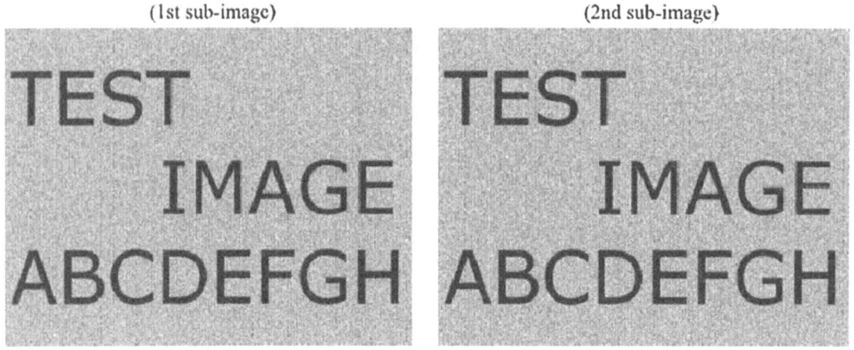

Fig. 4.17 Two sub-images produced by image add noise technology

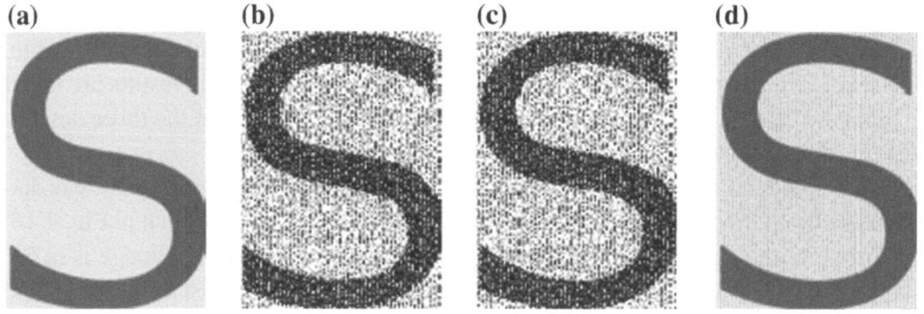

Fig. 4.18 The original noise adding, noise reduction, and average effect picture of letter "S"

is shown in Fig. 4.18b and c, respectively. It can be seen that the image becomes vague due to random noise adding. Figure 4.18d is the obtained image after averaging Fig. 4.18b, c.

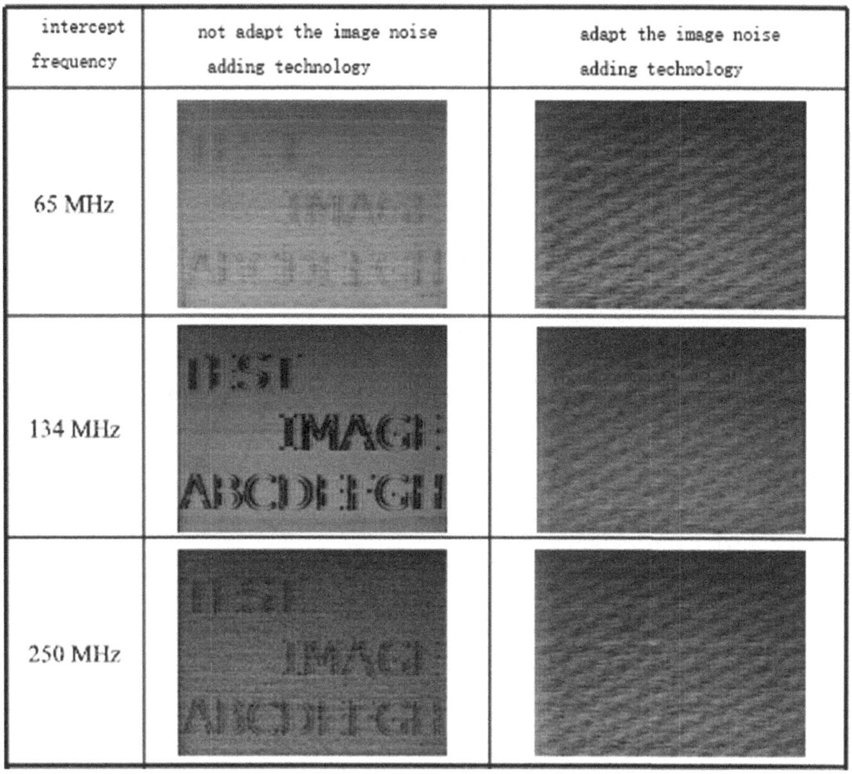

Fig. 4.19 Comparison between images in CRT before and after image noise adding

In the following, CRT and LCD are regarded as the target system, respectively, by which testing results are obtained.

Electromagnetic leakage result of CRT system is summarized in Fig. 4.19. In the second column of Fig. 4.19, without applying image noise adding, the strongest leakage of character edge is generated in the frequency range of 40–80 MHz, 120–180 MHz, 210–270 MHz, corresponding images are shown in Fig. 4.19. It is found that these frequencies are harmonic frequencies of XGA screen pixel clock frequency 65 MHz.

The third column in Fig. 4.19 gives the result by applying image noise adding. It is obvious that discernible character edge in the second column disappeared in the image of the third column, which can be explained as the use of image noise adding makes image reproduction impossible.

4. **LCD**

In this setup, CRT display is replaced by LCD display to do the same testing, whose results are shown in Fig. 4.20. Although the leakage is caused in lower frequency, the result is the same with Fig. 4.19 from the frequency range.

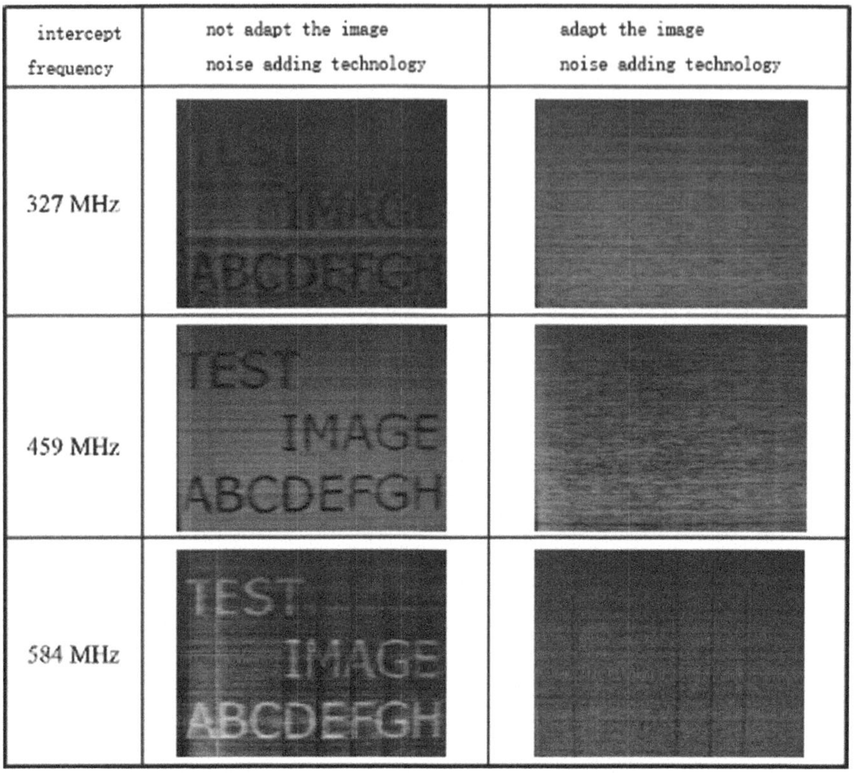

Fig. 4.20 Comparison between images in LCD before and after image noise adding

It is concluded that image noise adding can make image reproduction difficult if not possible with little loss of image quality. Although image contrast is decreased, it can be overcome by low-cost software. In addition, such a technique is independent with other electromagnetic protection methods, so which could be combined with other methods to realize effective protection of electromagnetic leakage.

4.3.4 TEMPEST Font Based on Fourier and Gaussian Methods

TEMPEST font based on Fourier and Gaussian methods is one of software techniques in the protection of electromagnetic information leakage; it is an effective protection technique improved by Gaussian filter on the basis of traditional TEMPEST font [44–46]. Compared with pure hardware protection measures of electromagnetic shielding, TEMPEST font of Fourier and Gaussian methods is more flexible with lower cost.

1. **Effectiveness and Defect Analysis of Basic TEMPEST Font**

Basic TEMPEST fond is formed by filtering off the high-frequency component of normal font, suppressing electromagnetic information leakage and making information reproduction difficult. In the process of transformation between digital and analog signal, high-frequency spectrum generates lots of wave peaks in the carrier wave, which can be easily distinguished from space noise and used as a clue to reproduce the target image. Therefore, an effective protection strategy of electromagnetic information leakage is obtained by removal of the high-frequency component in the radiation electromagnetic wave spectrum.

The essence of TEMPEST font is to use Fourier transform as a low-pass filter, removing 30% from the top of ordinary font horizontal spectrum, so that it is difficult to reproduce leaked electromagnetic radiation wave even though it is intercepted. TEMPEST font removes the high-frequency component of the font, however, the applied low-pass filter is limited in the range of 30% without too much loss or influencing display effects; meanwhile, the difficulty is increased to reproduce font information.

With developing the performance of interception antenna and improvement of interception device and technology, TEMPEST font exposes its main defects gradually, which makes the expected effectiveness of protection for electromagnetic information leakage can't be obtained.

The defects in existing TEMPEST font are mainly in two aspects:

(1) Edge jitter problem: noise block attached in TEMPEST font causes edge jitter phenomenon surrounded in characters. Such characteristics make it much easier to reproduce character information by "edge enhancement" filter and other technique.
(2) Easy to distinguish the electromagnetic spectrum of displayed character from noise. Because of a fine line constituted in electromagnetic leakage spectrum in displayed ordinary font, high-frequency spectrum generates a peak of carrier, which is easy to attenuate after radiation, so that it is difficult to distinguish from noise. While TEMPEST font deals high-frequency component, on the contrary, it is easier to distinguish from noise. That is to say, with the development of antenna's sensitivity and resolution and improvement of intercepting device and technology, reconstruction, and reproduction of TEMPEST font are much easier to achieve compared with the ordinary font.

2. **TEMPEST Font Based on Fourier–Gaussian Methods**

As the defect and limitation of TEMPEST font described above, TEMPEST font based on Fourier and Gaussian methods is introduced, which carry out Gaussian filtering on the base of Fourier transformed traditional TEMPEST font. The objective of it is to make image smoothing and keep higher correlation among adjacent pixels. After multiple filtering, two main problems are tackled.

(1) Noise block (jitter) phenomenon is solved and eliminated to raise the difficulty of interception and image reproduction.

(2) Decrease readability. Noise in TEMPEST font image is uniformly distributed in the whole background after Gaussian filtered, so that image readability is reduced after reconstruction and image reproduction. Meanwhile, it is difficult to easily distinguish the electromagnetic spectrum of displayed character from noise.

In summary, TEMPEST font based on Fourier and Gaussian methods greatly improves deficiency of traditional TEMPEST font, and it is an effective and reliable software-based electromagnetic information protection technique.

3. **Testing Examples**

The following describes examples in detail, which compare TEMPEST font based on Fourier and Gaussian methods with traditional TEMPEST font and prove that the former is more effective in protection of electromagnetic information leakage [45].

Test target is composed of three devices, which are an IBM ThinkPad S30 laptop and a SONY VAIO PCG-V505 laptop and a NANAO FlexScan 77F 21-in. CRT displayer, respectively, CRT is connected to VGA interface of SONY. Near-field probe is applied to receive electromagnetic radiation signal of the target device, followed by reconstruction and reproduction of original information.

Original TEMPEST font image is shown in Fig. 4.21; Fig. 4.22 shows reconstruction and reproduction font by IBM PC; Fig. 4.23 shows reconstruction and reproduction font by CRT connected with SONY PC.

Many characters are recognizable in two reproduction font images in Figs. 4.22 and 4.23, which means semantic content of traditional TEMPEST font text can be obtained. It is proved that the protection effect of electromagnetic information leakage by traditional TEMPEST font is unreliable.

The subjective comparison of reproduction effects with near-field probe, antenna, and injection probe. In which character recognition rate up to 80% is "readable", character recognition rate up to 50% is "basic readable" and without any discernible character is "unreadable". The data by antenna in Table 4.7 indicates that it is much easier to distinguish TEMPEST font from ordinary font after reproduction, which proves the limitation of TEMPEST font.

Fig. 4.21 Original image of TEMPEST font

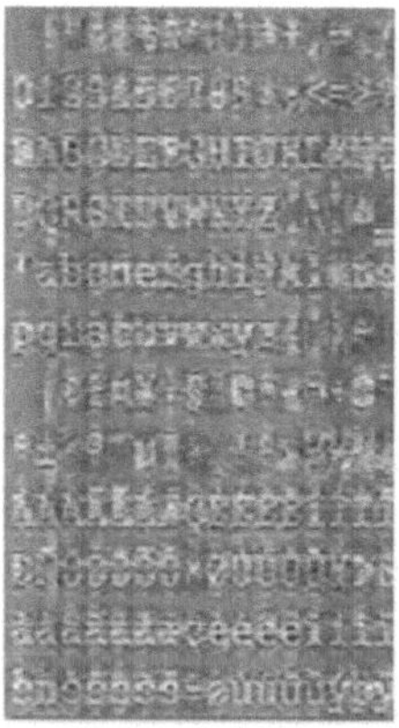

Fig. 4.22 Reconstruction and reproduction image of displayed font in laptop

Fig. 4.23 Reconstruction and reproduction image of displayed font in CRT

Table 4.7 Readability comparison after reproduction among near-field probe, antenna, and injection probe

	Near-field probe		Antenna		Injection probe	
	TEMPEST font	General font	TEMPEST font	General font	TEMPEST font	General font
VAIO	Readable	Readable	Readable	Unreadable	Unreadable	Unreadable
IBM	Readable	Readable	Unreadable	Unreadable	Unreadable	Unreadable
CRT	Readable	Readable	Readable	Unreadable	Readable	Readable

(a) TEMPEST font (b) Fourier-Gaussian method TEMPEST font

Fig. 4.24 Comparison of TEMPEST font and Fourier–Gaussian method TEMPEST font original image

Three kinds of intercepting devices are compared, which are antenna, injection probe, and near-field probe. Near-field probe gives the best interception effect of electromagnetic information leakage. So, if countermeasure verified ideal against reproduction of electromagnetic information leakage using a near-field probe, it can also obtain better protection effect against antenna and injection probe. The following reproduction image is derived from electromagnetic information intercepted by near-field probe, which can verify the effectiveness of TEMPEST font based on Fourier and Gaussian methods.

Figure 4.24 shows the comparison between original enlarged image of traditional TEMPEST font and TEMPEST font based on Fourier and Gaussian methods.

According to Fig. 4.24a, noise block (jitter) at the edge of characters can be recognized, from which characters can be reconstructed and reproduced by "edge enhancement" filter. So, existing noise block makes it easy to reproduce the font. Figure 4.24b is an enlarged image of TEMPEST font based on Fourier and Gaussian Methods.

TEMPEST font based on Fourier and Gaussian methods is made by multiple Gaussians filtering to eliminate noise block, the reproduced font can't be recognized, as shown in Fig. 4.25.

By statistical result from Fig. 4.25, the observer can recognize 80% of characters in TEMPEST font; however, nearly none of TEMPEST font based on Fourier and Gaussian methods can be discerned.

TEMPEST font based on Fourier and Gaussian methods is a soft protection strategy of electromagnetic information leakage, whose high-frequency component of font is eliminated by image smoothing; meanwhile, block noise is removed from the image, and thus, reproduction difficulty of font image is increased. Compared with traditional TEMPEST font, TEMPEST font based on Fourier and Gaussian methods greatly promotes the effectiveness of protection of electromagnetic information leakage.

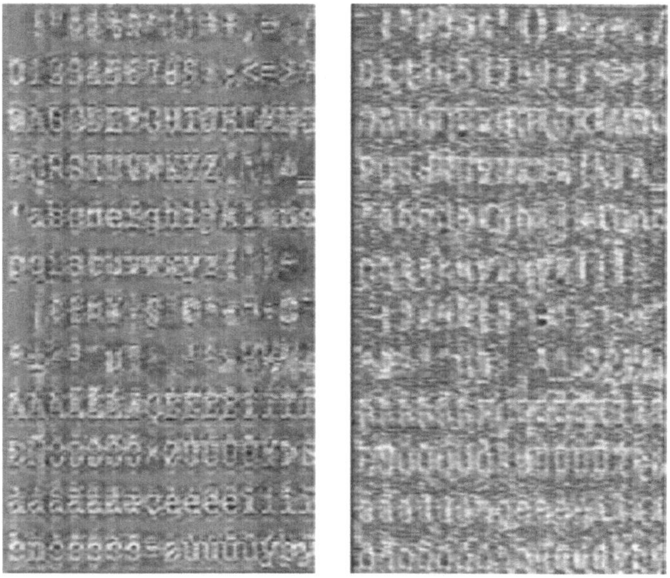

(a) TEMPEST font

(b) Fourier-Gaussian method TEMPEST font

Fig. 4.25 Comparison diagram of TEMPEST font and Fourier–Gaussian method TEMPEST font recurring effect

It should be noted that reproduction difficulty is increased by TEMPEST font based on Fourier and Gaussian methods, the reader will be influenced, too. That is the font will become dim and vague with the reduction of contrast, it is difficult to read the fine characters and symbols. So, the relationship between protection effectiveness and font visibility should be well balanced when TEMPEST font based on Fourier and Gaussian methods is adopted. The appropriate Fourier conversion and Gaussian filter parameters need to be carefully selected according to the type and size of the font and corresponding environment.

4.3.5 Custom Kernel Design

For data processing and transmission of the digital system, only part of data is very important for information security, which is called sensitive data. Sensitive data should be taken as protection emphasis of electromagnetic information leakage. For the design of chip kernel, if sensitive area and nonsensitive area are carefully divided, limiting sensitive data only in sensitive area, data security can be promoted. As far as the transmission of sensitive data is concerned, data security can be achieved by data hiding method.

1. **Design for secured processing of sensitive information**

In order to prevent sensitive information from leakage, protection measurements should be considered from time and space aspects. For the design of chip kernel, reduction of sensitive data processing unit and transmission path can decrease the possibility of electromagnetic leakage for sensitive data in space; the longer processing time for sensitive data, the larger possibility for it to be intercepted, the greater information security would be threatened. Shortening processing time can reduce the time of sensitive information leakage, so that the probability of interception is decreased.

Usually, the design of CPU is mainly composed of hardware design and instruction set design, they are tightly combined and jointly achieve CPU function. Hardware design mainly includes ALU (Arithmetic Logic Unit), register and so on; instruction set design comprised of instruction format, function command, and so on. The general working process of CPU is as follows: instruction fetching, sending operands to register for calculation, and result storing. Among which calculation of CPU is achieved by ALU, the calculation result is stored in memory or send into operand register directly or by general register and go on next step calculation, as shown in Fig. 4.26.

The general CPU shown in Fig. 4.26 is designed for general calculation. Sensitive data is involved in data processing by CPU, which requires higher security protection and is the main target of side channel attack [48,49], so it should be the protection emphasis of electromagnetic information leakage.

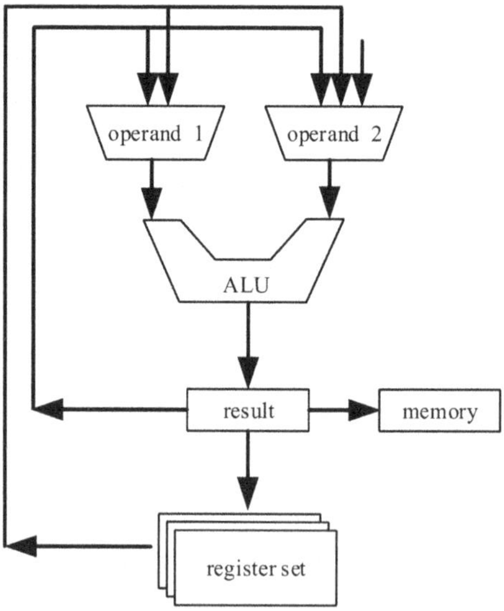

Fig. 4.26 Structure of general CPU

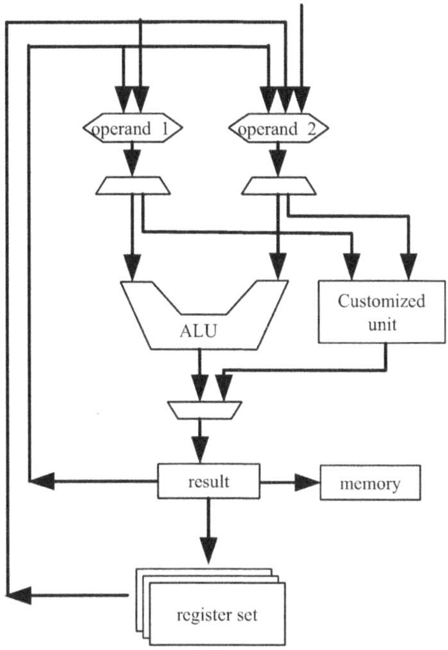

Fig. 4.27 Customized design of CPU

For the design of chip kernel, the strategy adopted is to distinguish between the custom computing unit and general computing unit, which are relatively independent. Sensitive data is calculated by custom computing unit, while others by general computing unit, which is shown in Fig. 4.27.

When custom computing unit is designed, the security logic design pattern is applied to achieve, CPU instruction set also is relevantly extended for the purpose of security rapid calculation.

The custom unit can improve the calculation power and obviously increase the calculation speed. The security of digital is also accordingly improved and decreased the attacked probability because the calculation time of the sensitive data is shortened.

In order to further improve the security of the sensitive data, on the base of such design, the improvement measures are also adopted to reduce the paths of electromagnetic information leakage.

The risk of electromagnetic information leakage is decreased by adding a splitter and let the sensitive data to transmit in a private path. The information leakage paths are reduced by adding internal dedicated registers set into CPU, avoiding frequent I/O between CPU and memory, so as to reduce leakage path of electromagnetic information, which is illustrated in Fig. 4.28; its main objective is to let sensitive information take solid-path not dotted path.

The security of sensitive data is strengthened by the improvement of custom design which is realized the proprietary of calculation unit of the sensitive data and

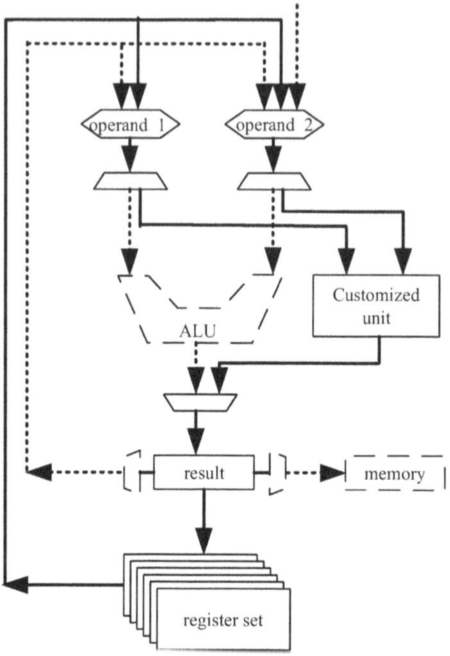

Fig. 4.28 Improved customized design of CPU

the transmission paths and limit the process area of the sensitive data and reduce the transmission paths and improve the calculation speed.

In a word, by adding dedicated registers set, sensitive data is split from other data and take private channel, so that sensitive data can be well protected.

For example, when AES encryption is performed in a general CPU, data is iterated by ALU unity for loop process. This process is vulnerable to attacks such as differential power analysis (DPA) and differential electromagnetic analysis (DEMA), resulting in electromagnetic information leakage. As described above, the process of loop iteration is executed in the added private unit, because the unit is independent and secure, in the meantime, calculation time is reduced. Therefore, the risk of electromagnetic information emanation is greatly reduced. The experiment result shows that, by using the added private unit, the calculation time of AES encryption is reduced to nearly one-tenth of the original time, and thus, the difficulty of deciphering the key is extremely increased.

2. **Information hiding transmission method**

Private unit help to improve information security of sensitive; information hiding in the transmission process can further ensure information security. Information hiding methods include hardware and software means.

(1) Masking method

As illustrated in Fig. 4.29, decoder and coder are added before and after private unit, respectively. That is to say, plain text only exists in the private unit.

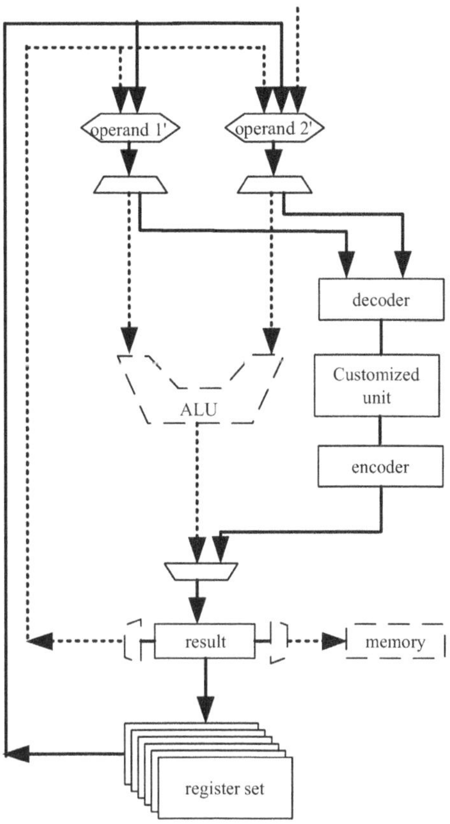

Fig. 4.29 Customized design of CPU with data hiding

Figure 4.30 is an example of decoder and encoder realized by the masking operation, whose essence is of XOR mask operation. Masking generator produces random numbers, which are stored in mask memory.

The principle of data hidden by mask operation is as follows:

$$X \oplus M = X' \tag{4.1}$$

$$X' \oplus M = X \tag{4.2}$$

Operand 1′ and operand 2′ are encoded operands, they are XORed in the decoder to obtain original operand 1 and operand 2, and then operand 1 and operand 2 are put into the customized unit to finish the calculation. The result of the calculation is encoded again and then output to the data transmission path.

(2) Inserting random number method

Inserting random number can also achieve information hiding. The location of the insertion needs to be studied carefully and does not affect the correctness of the

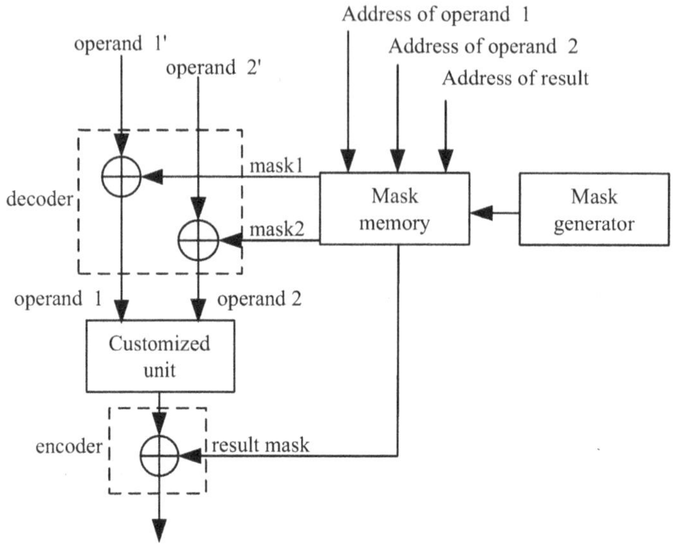

Fig. 4.30 Information hiding based on mask

original algorithm. Insert one or several random numbers at appropriate locations to change the voltage level on the data transmission path for achieving information hiding, as shown in Fig. 4.31.

For the digital circuit, high and low voltages are represented as bit "1" or "0", respectively. In order to show the change of voltage level, we adopt a two-bit description format, the first bit is the preamble status, the latter bit is the subsequent status. For example, "01" indicates the previous status is "0" and current status is "1", "00" shows previous status is "0" and current status is also "0", and so on. Among them, "01" and "10" indicate that the adjacent status changes, and "00" and "11" indicate that the adjacent status does not change. For eavesdropper, he or she needs to distinguish "0 → 0" or "0 → 1" to judge the current bit in the case where the previous status is "0". Data start with the status of "0" may be "00" or "01". For case "00", insert one bit between "00", the data could be "000" or "010". For case "01", after inserting one bit, the data could be "001" or "011", as illustrated in Fig. 4.31. Eavesdropper needs to distinguish among "000", "010", "001", and "011" to judge current status, so as to increase the difficulty of eavesdropping.

4.3.6 Photoelectric Isolation Method

Information device needs a power supply and communicates with the outside through data cable. Meanwhile, power line and data cable are two vital leakage sources of electromagnetic information. Photoelectric isolation method is implemented as follows: keep internal circuit unchanged, while replacing the power supply with the optical source and replacing the electric transmission with optical transmission, in

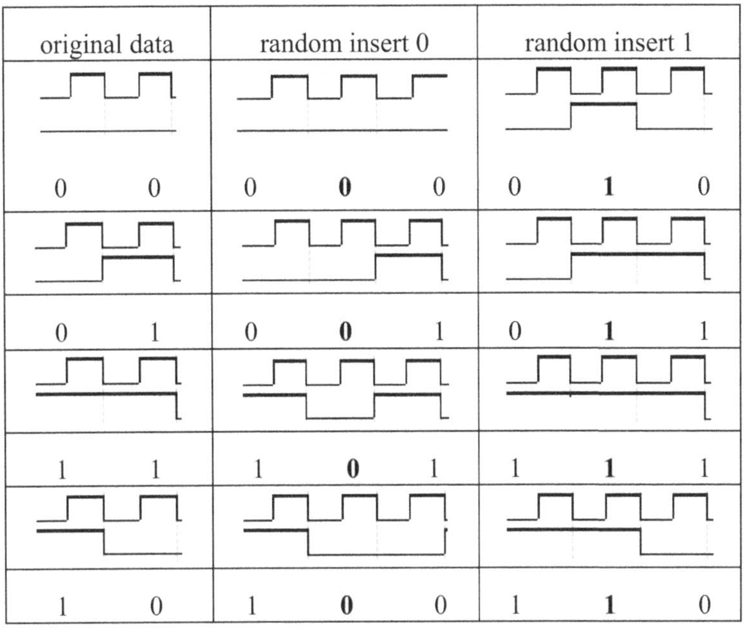

Fig. 4.31 Information hiding based on random number inserting

order to achieve external optical transmission, which can reduce the electromagnetic radiation phenomenon radically, so that information security is protected. The two methods above are explained as follows:

1. **Principal of photoelectric isolation**

Chips and the external circuit are connected by wire; electromagnetic radiation caused by chip computing is coupled into the external circuit by wire, capacitor, and resistor, which results in electromagnetic information leakage, by means of photoelectric isolation, that is, replacing the wire with optical lighting to block conduction coupling between the chip and the external circuit.

For reset circuit of MCU in Fig. 4.32, the electric signal generated by internal computing can be coupled into external pins. When performing an encryption operation by MCU, the electrical signal is coupled into reset pin, and periodic pulse signal can be detected by an oscilloscope, just as shown in Fig. 4.33. The periodic waveform corresponds to the electronic signal generated by the internal cyclic iterative operation.

Redesign the reset circuit of MCU by means of photoelectric isolation, as shown in Fig. 4.34; optical coupling component is introduced into the circuit. According to Fig. 4.35, periodic pulse signal can't be found by oscilloscope at position B, and its waveform is smooth and steady, which demonstrates that photoelectric isolation method can effectively block conduction coupling and greatly reduce electromagnetic information leakage.

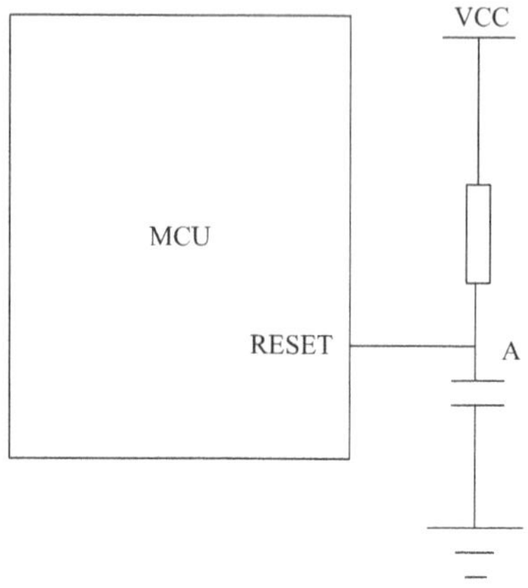

Fig. 4.32 Reset circuit of MCU

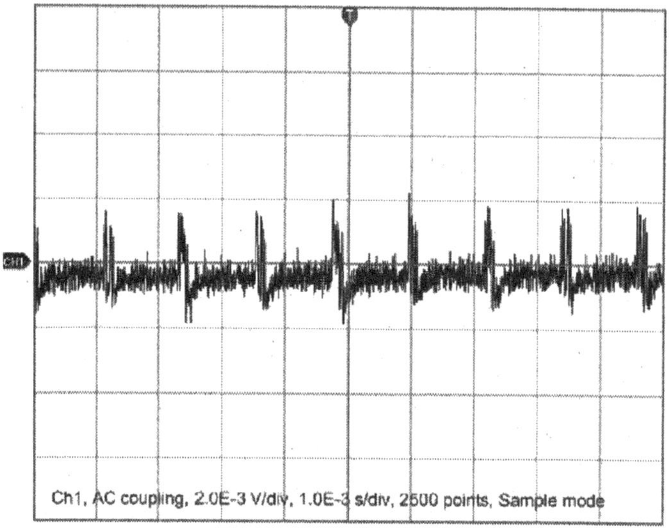

Fig. 4.33 Periodic pulse signal of point A in Fig. 4.32

2. **Photoelectric isolation implementation**

For circuit design, bidirectional photoelectric coupling unit can be applied to data transmission, thereby resulting in isolating conductive coupling of the electrical signal [50]. Bidirectional photoelectric coupling unit integrates two optocouplers to form a bidirectional optical transmission path of data, as shown in Fig. 4.36.

Bidirectional photoelectric coupling unit can isolate conduction coupling from the signal transmission path. However, power line in information device also produces conductive coupling, causing leakage of electromagnetic information. In order

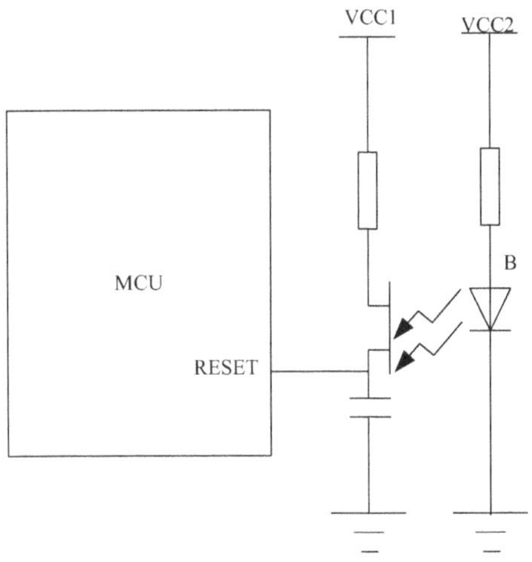

Fig. 4.34 Improved reset circuit of MCU

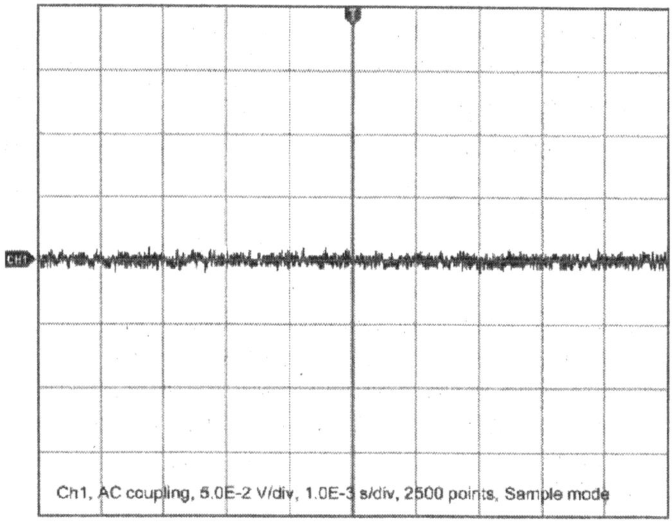

Fig. 4.35 Signal point B in Fig. 4.33

to solve this problem, the photovoltaic unit can be applied in internal power, and the internal electric circuit is driven by electric energy generated by photoelectric conversion, thereby can replace the external power supply. Thus, the connection between information device and external power cable is eliminated.

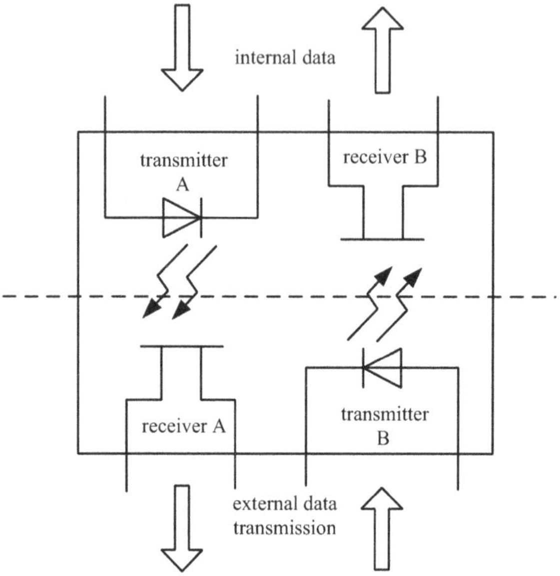

Fig. 4.36 Bidirectional photoelectric coupling unit

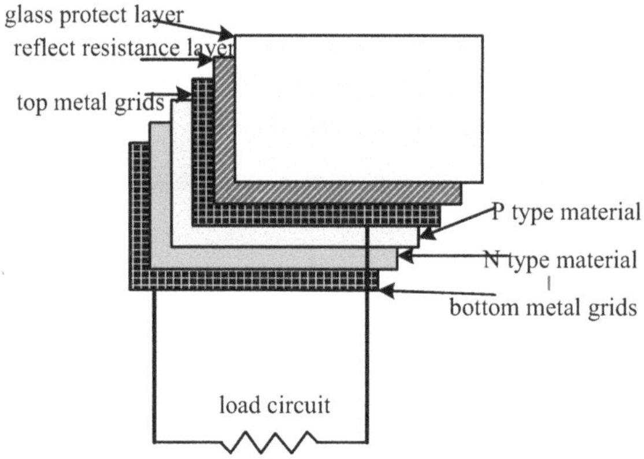

Fig. 4.37 Photovoltaic cell constructer chart

As shown in Fig. 4.37, photovoltaic cell is composed of six-layered material: the top is glass protected layer, layer 2 is reflection resistant layer, layer 3 and layer 6 are metal grids, layer 4 and layer 5 are P- and N-type materials, respectively, thus forming P–N junction. When P–N junction is illuminated, it forms electron–hole pair. Electric field generated by illumination on P–N junction, holes flow from P region to N region, forming a current flow to drive circuit.

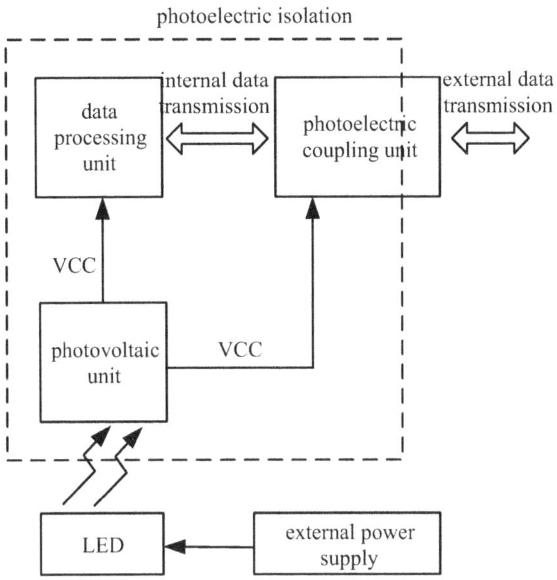

Fig. 4.38 Photoelectric isolation design

Figure 4.38 shows a photoelectric-isolated circuit with an integrated bidirectional photoelectric unit and photovoltaic unit. External power supply in the figure only drives the LED light source, with no connection between power and internal circuit.

The photovoltaic unit receives light from LED, converts light energy into electrical energy, and drives internal circuit. Internal signal interacts with outside in form of light through optocoupler unit.

The above design isolates conductive coupling of internal circuit from outside, but leakage of electromagnetic information caused by radiation coupling inside the circuit still exists. Therefore, electromagnetic shielding measure is required to shield the internal working circuit, such as the closed area drawn by a dotted line in Fig. 4.38. A good overall shielding effect can be achieved by comprehensively utilizing various protection technologies.

Chapter 5
Protection Material of Electromagnetic Information Leakage

Electromagnetic wave is a major way of electromagnetism energy transmission, which is radiated to outside when the high-frequency circuit operates. Since the electromagnetic waves radiated from the device are strongly related to the information being processed, they may cause information leakage and threat to information security.

By using anti-emanation material for restraining electromagnetic radiation, its objective is to solve the problems of electromagnetic information leakage and interference. In other words, the mechanism of anti-emanation material cuts the radiation path of the electromagnetic wave, so as to reduce or eliminate electromagnetic information leakage and interference.

5.1 Traditional Electromagnetic Shielding Material and Theory

5.1.1 Electromagnetic Shield Theory

Energy of electromagnetic emanation is transmitted by means of conductive coupling and radiation coupling. Electromagnetic shielding is the basic and most effective method among many solutions for electromagnetic information leakage. The advantage of shielding method is that it can keep the device's normal function without an interference. In order to prevent electromagnetic information leakage, shielding technique can be applied to suppress conductive coupling and radiation coupling.

The theory of the electromagnetic shielding is the shielding object that provides an electromagnetic pathway for the interference electromagnetic field with low impedance and low magnetic resistance, in which the far zone field of the radiation interference source is shielded to reflect and absorb the electromagnetic wave by the metal shielding object. Meanwhile, the component of the electric field and magnetic field caused by the field source can be shielded, as shown in Fig. 5.1. As a result,

T. Liu and Y. Li, *Electromagnetic Information Leakage and Countermeasure Technique*,
https://doi.org/10.1007/978-981-10-4352-9_5

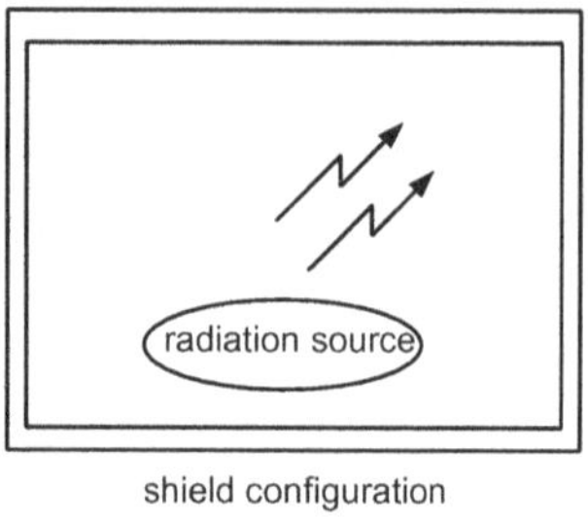

Fig. 5.1 Shielding sketch map

the electromagnetic information is separated from inference electromagnetic field. Therefore, it is very important to select a suitable material to make a closed shielding shell. Generally, materials with high conductivity and high magnetic permeability are selected, such as steel, iron, bismuth alloy, and metallized polymer materials, and so no. Electromagnetic shielding technology has a significant protective effect, in the complex electromagnetic environment, especially a large number of different levels of devices or devices work together, and the electromagnetic power density increases sharply within the unit volume.

5.1.2 Traditional Electromagnetic Shielding System

Traditional electromagnetic shielding system is classified into ten kinds of systems, which is included series of electromagnetic shielding window, series of electromagnetic shielding ventilation waveguide, series of electromagnetic shielding finger spring kinds of product, series of electromagnetic shielding metal screen gasket, series of electromagnetic shielding coating, series of electromagnetic shielding electroconductive rubber, series of electromagnetic shielding conductive resin and heat shrink tubing and shielding foil, and series of electromagnetic protection auxiliary thermal interface material.

According to different applications, various electromagnetic shielding materials can be chosen. Characters and limitations of traditional electromagnetic shielding material are summarized through the introduction of several series of electromagnetic material as follows.

1. Series of Electromagnetic Shielding Windows

Series products of electromagnetic shielding windows possess the functions of preventing the intra-system from electromagnetic information leakage and outside system from electromagnetic interference, which is applied in electromagnetic protection of visible part for graphic display terminal to meet the visually demands with higher transmittance. Special technology methods can be adopted to increase special function according to different applications, such as heating, bulletproof, flexible deformation, ultrathin, and so on.

Series of electromagnetic shielding window is included high-definition electromagnetic shielding glass, conductive glass, electrically heated glass, flexible shielding window, shielding film visible material, electronic security glass, and so on, as shown in Table 5.1.

The foundation technology index of series products for electromagnetic shielding window is included: electromagnetic shielding effectiveness, transmittance, temperature, and so on, in addition to the relative technology index, such as interface, thickness, heating voltage, insulation resistance, and so on, according to the special demands.

2. Series of Electromagnetic Shielding Ventilation Waveguide

The contradiction between ventilation and the shielding is solved by electromagnetic shielding ventilation waveguide in the electric facilities according to the theory of cutoff waveguide. It is similar to the frequency features of high-pass filter, in which the waveguide allows to pass the signal above the cutoff frequency and prevents or attenuates the signal below the cutoff frequency. The waveguide is called the cutoff waveguide to make the effect of electromagnetic shielding when the frequency of

Table 5.1 Series products of electromagnetic shielding window

Type	Technical indicators	Application
High-resolution electromagnetic shielding glass	Electromagnetic shielding effectiveness in the range of 15 kHz–10 GHz, up to 10–55 dB; transmittance ≥40 to ≥80%; operating temperature −45 to +65 °C	High-definition displays and precision instrumentation
Conductive glass	Transmittance ≥75%, flexible conductive pad connector, conventional thickness 0.5–5 mm	Precision electronic instrument display window
Electric heating glass	Interelectrode resistance 6–10 Ω; transmittance ≥85%; operating temperature −55 to +70 °C; insulation resistance ≥100 MΩ	Increase the local ambient temperature of the monitor in cold and humid environments
Flexible shield window	Electromagnetic shielding effectiveness range 15 kHz–10 GHz, 10–55 dB; light transmittance ≥65%; operating temperature −45 to +65 °C	Complex shapes, slim displays, and precision instrumentation
Shielding film visible material	Lightweight and flexible	Electromagnetic protection visible windows and special-shaped electromagnetic protective window
Electronic security glass	Meet special requirements	Bulletproof, all-weather use requirements

the interference signals is designed in the cutoff range of waveguide according to such feature.

Main features of electromagnetic shielding ventilation waveguide are better ventilation characteristics and higher shielding features and larger of demands for various electronic systems in military and civil field, such as anechoic chamber, shielding room, vehicle equipment, airborne equipment, cabinet, case, and so no. Now, there are three kinds of series which are classed into general electromagnetic shielding ventilation waveguide and light types of electromagnetic shielding ventilation waveguide and foam metal type of electromagnetic shielding ventilation waveguide at home according to the different applications as shown in Fig. 5.2. The technical specification is listed in Table 5.2.

3. Product series of electromagnetic shielding finger spring kinds

Electromagnetic shielding finger spring is electromagnetic protection series products which is of high elasticity and high electroconductivity and reusable, shielding effectiveness can be finished in the on–off facilities no matter the door and window parts termination connection part and side aperture part, the products can be installed to apply in any occasion of sliding friction.

Various coatings can be chosen to ensure electrochemical compatibility with other connection surface and without generating a spark.

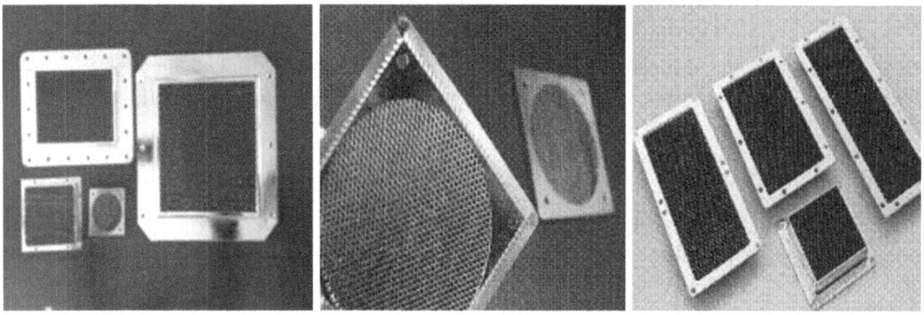

Fig. 5.2 Series products of ventilation waveguide

Table 5.2 Technology specification of series products of ventilation waveguide

Type	Technical indicators
Universal electromagnetic shielding ventilation waveguide	Shielding performance 14 kHz–40 GHz, 60–90 dB Operating temperature −55 to +75 °C Wind resistance ≤8
Lightweight electromagnetic shielding ventilation waveguide	Shielding performance 45 MHz–18 GHz, ≥60 dB Operating temperature −55 to +75 °C Wind resistance ≤8
Foam metal electromagnetic shielding ventilation waveguide	Shielding performance 14 kHz–2.5 GHz, ≥60 dB Operating temperature −55 to +75 °C Wind resistance ≤15

Fig. 5.3 Classification of finger spring product

- Spring
  - Heterogeneous type
    - Bent type
    - Semicircular opening type
    - Closed semicircular type
    - Right angle type
  - Sawtooth type
    - Single Sawtooth type
    - Double Sawtooth type
    - Positive sawtooth type
    - Inverted sawtooth type
  - Riveted type
  - Connector type

Product series of electromagnetic shielding finger spring kinds is suitable for electromagnetic shielding for various types of electronic facilities and different military electronic equipment and shielding room because of various types and many styles of installation. Serial products can be divided into four kinds, which are heterotypic construction finger spring, primordial type of finger spring, connection type of finger spring, connector sheep type of finger, shown in Fig. 5.3.

Characters of electromagnetic shielding finger spring are higher elasticity and a wider range of deformation and better environment suitability but poor hydrosphere sealing; the products are shown in Fig. 5.4.

General technology performance index is given in the following: when shielding effectiveness is 10 kHz–30 GHz, no more than 50 dB, environment adaption temperature is −65 to +125 °C.

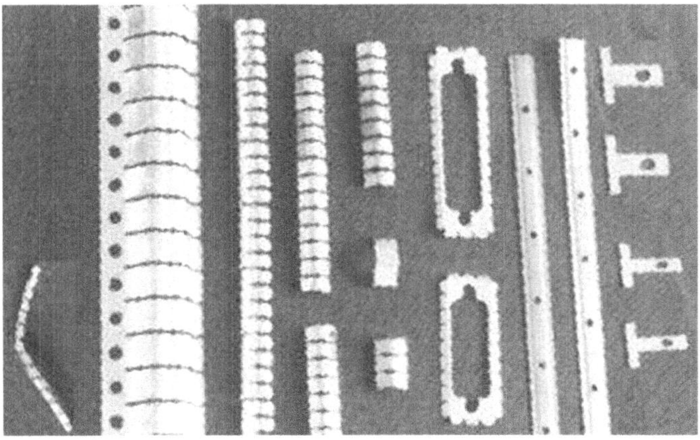

Fig. 5.4 Products of finger spring kinds

4. Series of Electromagnetic Shielding Metal Screen Gasket

Electromagnetic shielding metal mesh gasket is a kind of electromagnetic shielding material which is applied in the seams of electronic equipment frame and the door and the window of the shelter and cabinet and case. It not only prevents interference from the outside magnetic field, but also avoids the leakage of inside confidentiality information and ensures the environmental sealing.

Electromagnetic shielding metal mesh gasket is classed into two kinds: one is mesh gasket with an elastic core, and another is mesh gasket with all-metal. The structure of the shielding gasket with the elastic core is combined with the waving wire mesh of good elastic and electroconductivity with the elastic material of the features of better compression and deformation together. All-metal mesh gasket is the gasket mesh strip waved by metal wire with better electroconductivity and corrosion resistance; commonly used materials in Monel metal are: Su Ph Bz, and beryllium copper wire. Main technology performance index: shielding effectiveness is 10 kHz–30 GHz, 40–95 dB, working temperature is −55 to +125 °C. Maximum compressive deformation ≥50%, compression permanent deformation <15%, the product of metal wire shielding gasket is shown in Fig. 5.5.

5. Series of electromagnetic shielding coating

Electromagnetic shielding coating is mainly a solution of electromagnetic protection for different kinds of electromagnetic information equipment and electronic facilities and security facilities. The products can be classed into polymer, silicone, polyenes, epoxy resin, and so on, according to the different base material and different products adhesive system, in which polymer-type electromagnetic shielding coating is suitable for macromolecule base material and other nonmetals, and silicone-type electromagnetic shielding coating is suitable for electrical connection of normal silicone rubber gasket surface. Polysene-type electromagnetic shielding coating is suitable for electrical connection of polyolefin class rubber gasket (such as EPDM). Epoxy resin-type electromagnetic shielding coating is suitable for the electromagnetic shielding in the environment of high-temperature resistance.

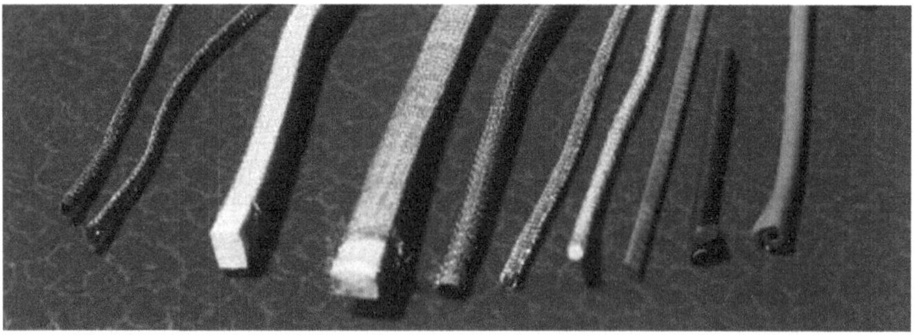

Fig. 5.5 Electromagnetic shielding metal wire gasket

The function characters of electromagnetic shielding coating are as follows:

(1) Surface metallization of nonmetal material.
(2) Prevent the interference from the outside electromagnetic field and leakage of inside electromagnetic information.
(3) Composition the coating layer with macromolecule elastomer with the function of environmental sealing.

6. Series of electromagnetic conductive rubber

The conductive rubber is a silicone rubber or a fluorosilicone rubber containing uniformly distributed conductive particles, such as silver-plated aluminum, silver-plated glass, silver-plated copper, nickel-plated graphite, silver, carbon. The conductive particles are brought into contact by pressure to achieve good electrical conductivity. Normally, the electronic conduction particle is included the aluminum silver plating and glass silver-plated and copper silver-plated and fine silver and graphite nickel-plated and carbon black, and so on.

Conductive rubber is classed into plate type and bar type and hollow type and abnormal type and the composition type according to the product's appearance, shown in Fig. 5.6. The different specifications of conductive rubber can be chosen according to different occasions; the features and occasion of various kinds of conductive rubber are shown in Table 5.3.

Fig. 5.6 Conductive rubber product

Table 5.3 Products of conductive rubber

Type	Features and applicable occasions
Aluminum silver plating series	Good electromagnetic shielding performance, electrochemical compatibility of aluminum materials, resistance to EMP shock. Recommended for high-end products aluminum alloy chassis material
Glass silver plating series	Recommended for commercial grade products
Copper-plated series	Good electromagnetic shielding performance, resistance to EMP shock. Recommended for advanced products
Pure silver series	Prevent mold, recommended for microbial growth environment
Graphite nickel plating series	Lower price, electrochemically compatible with aluminum, recommended for general products

According to the traditional electromagnetic shielding theory, the better the electrical conductivity and magnetic permeability, the higher the shielding effectiveness. But, actually, it is impossible both of these feathers for the metal material, such as the electrical conductivity of the copper is better, however, the magnetic permeability is poor. Another one, the magnetic permeability of the iron is better, but the electrical conductivity is poor. When the shielding material is selected to focus on either electrical conductivity or magnetic permeability, the shielding is depended on reflection loss or absorption loss. In brief, the traditional shielding principle is mainly depended on the feature of the electrical conductivity and magnetic permeability; in other words, it mainly depends on the feature of the materials.

5.1.3 Integrated Application Case of Electromagnetic Shielding Material

Electromagnetic shielding cabin is an integrated application of technology of shielding and containment and source suppression. According to the published reports, the design of electromagnetic shielding cabin mainly fuses on the magnetic field with the low-frequency band and electromagnetic shield with the high-frequency band, which the shielding performance mainly dependent on the technology measurement adopted in the location of electromagnetic leakage of the shelter [51]. The shelter object is shown in Fig. 5.7.

The structural design of the shelter ensures the electrical continuity of the door the window, orifice, and so on. The model design of the shelter is including the following:

Fig. 5.7 Electromagnetic shielding cabin

(1) Choice of the wallboard material of the shelter

The wallboard material of the shelter is usually accordingly chosen the conductive material and permeability material according to the actual environment. The shielding performance of the aluminum sheet is better than others compared with each other which can be made of the inner and outer skin of the shelter. Tin-plated copper tape is adopted in the seam of the wallboard due to the better features in high and low frequency in order to increase the shielding effectiveness.

(2) Cabin door design

Cabin door is the largest leakage of electromagnetic shielding cabin and one of the most different designs; there are two reasons: the first reason is that the shielding effectiveness of the cabin door is heavily influenced by the frequency on–off state, the compression fatigue resistance performance, the electrochemical corrosion resistance of the gasket and conductive treatment of the door frame, wear-resisting property of the conducting layer, and so on. The second reason is that it is difficult to keep the conductive gasket of the door in a reasonably compressed state, which involves assembly, manufacturing tolerances and other waterproof requirements. So, the epoxy or silicon grease conducting adhesive with low electrical resistivity is added and a conductive shielding belt is pasted on the interface.

(3) Window design

The window is made of electromagnetic shielding glass, which has good light transmittance and high resolution, and can effectively suppress electromagnetic signals passing through the window and prevent information leakage.

(4) Ventilation design

Electromagnetic waves radiating from the vents are one of the important reasons for the decline in shielding effectiveness of the cabin. At present, the cut-off waveguide venting window is generally adopted, so that the square cabin can achieve the desired shielding performance and ensure good ventilation. The interconnection structure between the cutoff waveguide ventilation window and cabin wall is the design key which is correctly chosen according to the frequency range. The purpose of suppression electromagnetic leakage can be realized by increasing the thickness and decreasing the aperture and decreasing the area of single ventilation opening besides the seamless electrical connection between the ventilation and cabin wall.

(5) Wall tube design of air conditioner via hole

Copper pipes are used for making air conditioner via hole, and special finger springs are used to make the copper pipes close to the bulkheads.

(6) Horn coupling design

The horn coupling is applied the conductive socket and the conductive gasket included conductive rubber is added between the socket and adapter plate.

5.2 New Electromagnetic Shielding Material—Photonic Crystal

Photonic crystal is a kind of artificial crystal which is originally generated from the traditional crystal concept. The particularity of new electromagnetic protection material is possessed but without that of the traditional material.

5.2.1 Electromagnetic Feature and Advantage of Photonic Crystal

The material features of photonic crystal are the revolutionary change in electromagnetic protection products and technology.

1. Feature of Photonic Crystal

(1) Photon forbidden band feature of the photonic crystal.
Periodic potential field generated by the crystal structure of atomic arrangement will influence the movement behavior of the electron in the semiconductor material, and the electron will form the energy band structure. Meanwhile, it is possible to exit the bandgap between the energy bands. The influence caused to the photon is similar to that of the electron due to the dielectric constant of the periodic arrangement in the space, the photon bandgap structure is formed which is the photon forbidden band. Electromagnetic wave can't be transmitted to photonic crystal alone any direction in the forbidden band frequency domain, but the electromagnetic located in the photon energy band frequency domain can be loss-free transmitted which is the photon permit band.

(2) Photon localization feature of the photon crystal.
The phenomenon is called Anderson localization of electron wave which is from conductive to insulation caused by the impurity in the conductor. The photon localization is the defect mode or localization state (Anderson localization) with extremely narrow frequency band appeared in the forbidden band as possible after the impurity is added into the photon crystal and the original symmetry is destroyed.

The feature of the localization state is depended on the parameter of the defect, and the photonic crystal possesses the point defect and line defect. For point defect, the light can be captured in the certain area in which the light can't transmit to the outside in any direction and equal to be totally packaged by all of the reflection walls, for the line defect, the light located in the bandgap of the photonic crystal will be localized in the place of the line defect, which is only transmitted by the direction of the line defect and equal to a non-loss waveguide.

2. Electromagnetic Shielding Feature of Photonic Crystal

Traditional shielding theory is mainly depended on the material feature. As a new type of electromagnetic protection material, photonic crystal posses the features without in the traditional material to break the feature limits. The shielding effectiveness of the photonic crystal is mainly depended on the energy band feature of the photonic crystal. The electromagnetic wave of the forbidden band frequency domain is totally shielded in the photonic crystal, but the electromagnetic wave of the passing band is totally passed, so the good shielding effectiveness can be achieved by the forbidden effect of the photonic crystal.

The electromagnetic wave area corresponding to "forbidden band" and "passing band" is depended on the construction of the photonic crystal, which the frequency domain corresponding to the forbidden band can be adjusted by the construction design of the photonic crystal, so the frequency band of the electromagnetic shielding can be achieved to controlled effectively.

Photonic crystal is applied in design and preparation of electromagnetic shielding material, due to the feature of the designable construction and the adjustable bandgap; many limitations of the traditional material can be effectively solved to realize the performance of thin film and wider frequency and higher shielding effectiveness of electromagnetic shielding. Especially, in terms of visualization type of electromagnetic protection, the contradictory issue between high shielding effectiveness and high transmittance can be solved. So, the photonic crystal has tremendous application potential in the electromagnetic protection field.

3. Advantage of Photonic Crystal of Electromagnetic Shielding Material

Photonic crystal overture is the traditional shielding mechanism; it is possible to make the electromagnetic shielding technology revolutionary change. The new solution will be found in which the issue can't be solved by many traditional materials. The following problems can be solved by photonic crystal.

(1) Broadband

Traditional material mainly depends on the features of electrical conductivity and magnetic conductivity. In general, the magnetic shielding is the priority in the low-frequency band (10–500 kHz), and the electric shielding is focused in the high-frequency band (500 MHz–40 GHz). In general, materials with good magnetic permeability have poor electrical conductivity, and it is difficult to achieve both. Therefore, electromagnetic shielding of a wide frequency band is difficult to achieve. Due to the physics feature of the photonic crystal, the heterojunction photonic crystal will be formed to expand the width of the forbidden band.

(2) Shielding wave band

In actual application, the shielding of some wave band can be realized and highly passed in the adjacent band by control the structure of the photonic crystal, which "random tailor" of electromagnetic shielding wave band isn't obtained by the traditional materials.

(3) Lightweight

The traditional material has mainly applied the material with good electric conductivity and magnetic conductivity, which has larger density. The lightweight technology is up to the limits with technology development. Photonic crystal shielding materials break through the constraints of conductive and magnetic properties of materials, making it easier to select shielding materials. In a certain extent, the electromagnetic shielding material of photonic crystal is the thin-film material with periodic construction for which the quality almost can be ignored. It not only ensures the good shielding performance but also meets the needs of lightweight.

(4) Transparency of metal film

In traditional material, electric conductivity and visibility are restricted each other. The better the electric conductivity, the poorer the visibility. So, all of the metal with good conductivity is nontransparent. However, the energy band theory of photonic crystal overcomes the technical difficulty of conductivity and visibility of the metal so that it is opened up a new way for the electromagnetic shielding material used in visible section.

(5) Electromagnetic shielding of high integration density

With the development of electronic technology, electronic devices are getting smaller and smaller, and number of circuits integrated inside the chip is getting higher and higher. Better electromagnetic compatibility performance is required. Conventional metal shielding materials can easily raise its temperature due to their good absorption of infrared rays. This may affect the normal operation of the electronic device to a certain extent. The photonic crystal has good shielding performance and at the same time has good reflection ability to the infrared, thereby avoiding the influence of its thermal effect on the operation of the electronic device.

In general, the traditional shielding technology is constantly excavated and trend to the limits. But the photonic crystal is opened up new design space for the shielding technology. With the development of the photonic crystal, it is possible to push a new generation shielding material to born.

5.2.2 Electromagnetic Wave Shielding Material of Photonic Crystal

According to the traditional mechanism of the electromagnetic shielding, it can't be simultaneously realized high shielding performance and high transmittance. Traditional electromagnetic visible material is classified two kinds of metal mesh and metal film. There is an insurmountable defect. The mesh type of shielding windows with higher shielding performance is poor visible features which will influence to observe the graphic, and it is difficult to meet the needs of the window of electric facilities. However, the film type of shielding windows with better visible characters is of lower electromagnetic shielding performance. Due to the influence of electromagnetic shielding mechanism, there isn't a kind of material combined good shielding performance with good visible together. So far, it is difficult to break through in traditional electromagnetic shielding visible material.

On the base of the energy band theory of photonic crystal, the periodicity and film proportion of the photonic crystals are adjusted, the visible light is passband with good visible performance, and the shielding band is forbidden band with good shielding performance. So, the compatibility is available of the electromagnetic shielding and visible feature, and the technical difficulty solves the restriction of the electromagnetic shielding and visualization.

1. Electromagnetic shielding performance of photonic crystal film

The shielding performance of photonic crystal film is as same as the metal film through the experience, as shown in Fig. 5.8.

2. Visible performance of photonic crystal film

Although the total metal film thickness (40 nm) is much larger than the skin depth (5–7 nm) of visible light, the visibility still keeps good status in photonic crystal film, as shown in Fig. 5.9.

Therefore, electromagnetic shielding mechanism of photonic crystal overturns the traditional electromagnetic shielding mechanism.

The technology bottleneck of restriction between the shielding and the visibility can be solved by controlling the construction of photonic crystal. It is possible that the new mechanism drives the leapfrog development of visible material of electromagnetic shielding.

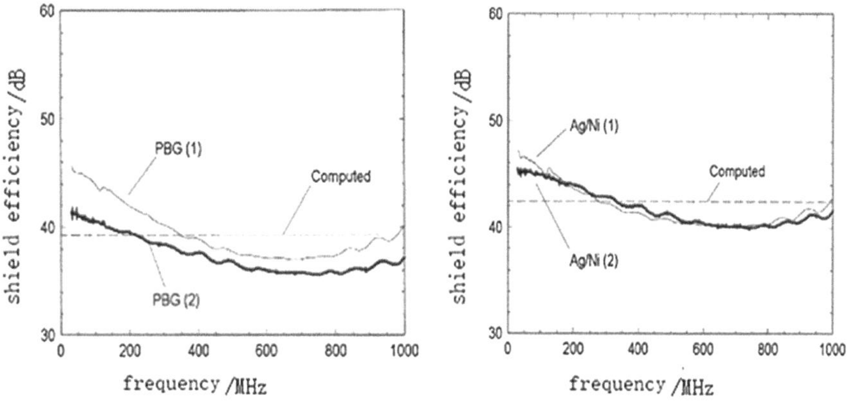

(a) PBG electromagnet shield performance (b) Ag/Ni film shield performance

Fig. 5.8 Comparison diagram of Ag film and PBG electromagnet shield which contains Ag in same thickness

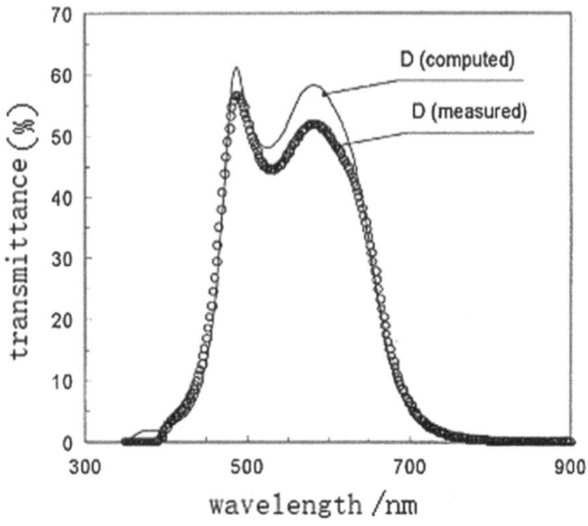

Fig. 5.9 Spectrogram of photonic crystal film

5.2.3 Electromagnetic Wave Absorbing Material of Photonic Crystal

Absorbing material is a kind of material which can absorb the electromagnetic wave energy projected on its surface and convert the loss of material into the thermal energy. The material requires higher absorptivity of electromagnetic wave in a wider frequency band in the engineer application, besides the performance of lightweight, moisture resistance, temperature resistance, corrosion resistance, and so on. In the general case, the absorbing material should possess two basic features.

(1) The incident electromagnetic wave comes into the internal of absorbing material at a maximum level in order to decrease the direct reflection of the electromagnetic wave which is impedance matching features.
(2) The material generates effective absorbing or attenuation to the incident electromagnetic wave which generates the electromagnetic loss to convert the electromagnetic wave energy into thermal energy or other energy in order to electromagnetic wave absorbed in the media at maximum level, which is the attenuation feature.

Photonic crystal is also called electromagnetic bandgap construction and microwave photonic crystal in the field of microwave and millimeter wave. The photonic crystal presents novel electromagnetic features due to the electromagnetic bandgap. The absorbing wave characters can be increased by applying the photonic crystal theory into the design of the absorbing wave material, which has great potential in the application of absorbing material.

1. Classical Salisbury Screen

The most basic construction of absorbing material is Salisbury screen, which is composed of three parts: impedance layer, media layer, and substrate. The absorbing wave of Salisbury screen is realized by setting the thickness to obtain the equal amplitude and opposite phase of reflection wave between the impedance layer and substrate for the purpose of interference cancelation. The thickness of the media layer must be 1/4 of wave. So, the thickness of absorbing construction for Salisbury screen is relatively larger and bandwidth is narrower.

2. Improvement of Salisbury Screen on Photonic Crystal

The impedance layer of Salisbury screen can be realized by the absorbing structure of electromagnetic bandgap. The periodic characteristics of bandgap construction of photonic crystal can be effectively attenuated the incident wave and absorbed the reflection wave came into the material, which is above two times of Salisbury screen of effective bandwidth and improves the bandwidth performance of the absorbing structure.

In fact, the photonic bandgap structure not only improves the absorbing properties of the Salisbury screen, but also breaks the thickness limitations of the Salisbury screen. By designing the photonic bandgap structure of the high-impedance surface, the thickness of the dielectric layer is no longer required to be 1/4 wavelength by the loss effect of the high-resistance layer, which reduces the difficulty of material production.

Common electromagnetic bandgap construction is as follows.

(1) High Impedance Surface, HIS
D. Sievenpiper and E. Yablonovitch pointed out a kind of superior electromagnetic bandgap which is called as High Impedance Surface (HIS). This kind of construction is made up of the array of series of metal protrusion attached on the surface of the metal plate. These metal patch arrays are connected to metal hole with substrate conductive plate, as shown in Fig. 5.10. This kind of construction is obtained by printed electric circuit technology.

The electromagnetic characters of the protrusion unit are described by the lumped parameter circuit if the size of the protrusion unit is far less than the wavelength. It is regarded as the paralleling LC syntonic circuit, the metal edge between two adjacent units is formed a capacity C, the longer conductive path is formed inductance L, and finally, the two-dimensional electric filter is formed and cut off the specific frequency current to flow on the surface of metal plate. When the incident wave frequency is equal to the resonance frequency, the synthesis electric field on the surface is the maximum tangential electric field which can be sufficiently consumed in loss of electronic screen. When the surface reflection coefficient of absorption screen is minimized, the absorption capacity of the screen is maximized, and the structure at this time has a characteristic of high impedance. The electromagnetic features of high impedance surface are changed by suitably improving the shape of syntonic unit and geometry parameter, and so on. The different absorbing bands can be conveniently

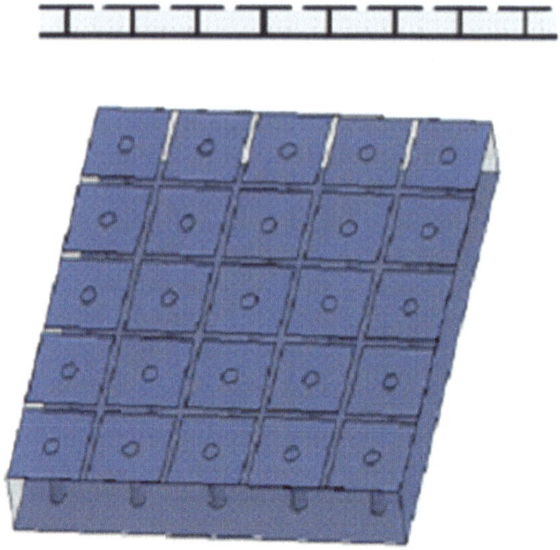

Fig. 5.10 HIS construction pointed out by Sievenpiper

obtained. The preparation is complex because the high impedance electromagnetic surface should be holed.

(2) Uniplanar Compact PBG, UC-PBG
Qian Yongxi and T. Itoh pointed out another uniplanar compact PBG construction which is a new style electromagnetic bandgap. This structure is shown in Fig. 5.11. It does not need to be connected between the metal piece and the substrate through the metal hole, making it easier to produce.

As shown in Fig. 5.11, the metal belt is connected between the adjacent unit which is provided additional inductance and the additional capacitance provided by the gap between units. The propagation coefficient of electromagnetic wave can be enlarged by this additional capacitance and inductance which can be obtained the

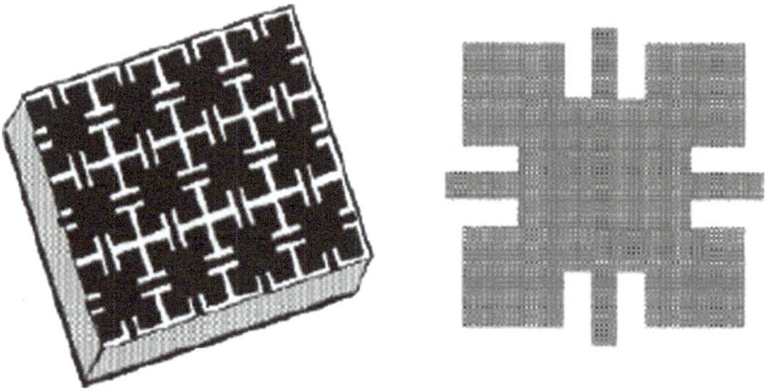

Fig. 5.11 UC-PBG structure planar graph

slow wave effect. Meanwhile, this kind of construction can obtain high impedance surface features. The resistance and the capacitance can be changed by adjustment and optimization of geometry shape of the aperture between the metal tape and adjacent unit. The frequency of the attenuation band can be further decreased and obtained a more contact construction. The enough larger stopband attenuation and bandgap width can be obtained to meet the needs of system performance by the design of UC-PBG unit construction in the condition of reducing the unit size as far as possible.

3. Relevance Design Theory and Analysis Tool

Fractal theory is applied when UC-PBG construction unit is designed. The fractal construction is normally generated self similar iteration of the initial unit by the definitive scaling factor; the initial unit determines the frame of the fractal pictures, and the iteration unit stipulates the construction of the fractal picture, such as Minkowski fractal.

The construction process of Minkowski fractal boundary is shown in Fig. 5.12. The initial "square patch" is marked as M0, its side length is L, and a rectangular area is dug from the intermediate position on its four sides, and the rectangle has a width of L/3 and a depth of h. The ratio of the depth to the width of the rectangular notch is p, that is, $p = 3\,h/L$ $(0 < p < 1)$ indicates the relative size of the notch depth. By changing the p-value, we can get different 1st-order Minkowski fractals. For all the straight edges of the 1st order Minkowski fractal, divide it by the same p value and get the 2nd order Minkowski fractal, as shown in Fig. 5.12c.

Fractal theory is suitable for solving the problems of scattering of coarse construction and wave propagation of inhomogeneous medium. Geometric of metal grid cell can be optimized selection by simulation software HFSS on the analysis of matrix unit. HFSS uses periodic boundary conditions and tetrahedral nonlinear mesh technology to analyze various arrays such as equilateral triangles, isosceles triangles, arbitrary triangles, hexagonal arrays, Y-shaped arrays, and rectangular arrays.

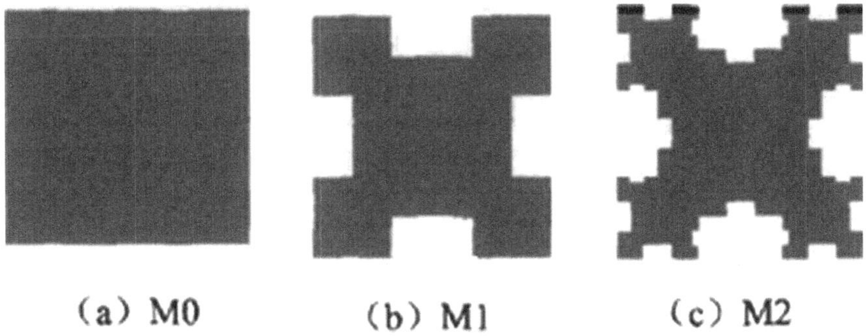

Fig. 5.12 Iteration generated process of Minkowski fractal patch

5.3 Other Electromagnetic New Material

With the development of material science, a more and more new material used in electromagnetic information leakage protection is appeared, such as left-handed materials, carbon fiber tube, and so on [52]. New materials are of unique advantage in the field of electromagnetic information leakage protection and more and more applied in the future.

5.3.1 Left-Handed Materials

Permittivity ε and permeability μ are two most basic physical quantities which describe the electromagnetic wave propagation characters in homogeneous media. Theoretically, the material is classified into four kinds according to the value of ε and μ: when $\varepsilon > 0$, $\mu > 0$, which is normal material, that is right-handed material; when $\varepsilon > 0$, $\mu < 0$, which is negative permeability material; when $\varepsilon < 0$, $\mu > 0$, which is plasma and metal. When $\varepsilon < 0$, $\mu < 0$, which is left-handed material, as shown in Fig. 5.13. Left-handed material is a kind of new material with special electromagnetic features.

It is seen from Fig. 5.13, in the first quadrant, electric field E and magnetic field H and wave vector K of electromagnetic wave of right-handed materials are formed right-handed helix relation. The direction is same between K and energy flux density. But in the third quadrant, electric field E and magnetic field H and wave vector K of

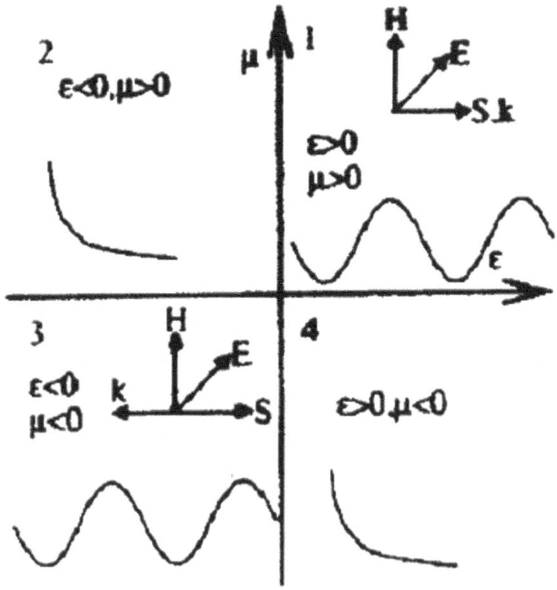

Fig. 5.13 Electromagnetic wave features in different media

electromagnetic wave of left-handed materials are formed left-handed helix relation. The direction is opposite between K and energy flux density.

The relation of the reflection and the reflection of right-handed material is shown in Fig. 5.14. When electromagnetic wave incidents to the right-handed material interface from the air, its reflection rays and the incidence rays are located in both sides of the normal.

The relation of the reflection and the reflection of left-handed material is shown in Fig. 5.15. When electromagnetic wave incidents to the left-handed material interface from the air, its reflection rays and the incidence rays are located in the same side of the normal, not in both sides of the normal.

Conventional electromagnetic wave absorbing materials are made of right-handed materials. In order to achieve the best absorbing effect, it is necessary to select appropriate electromagnetic parameters according to transmission loss of electromagnetic waves in the dielectric layer. This allows the input impedance of the dielectric layer to match the wave impedance of free space. Normally, double layers absorbing material construction is applied in order to obtain optimum matching of the electromagnetic parameter, as shown in Fig. 5.16.

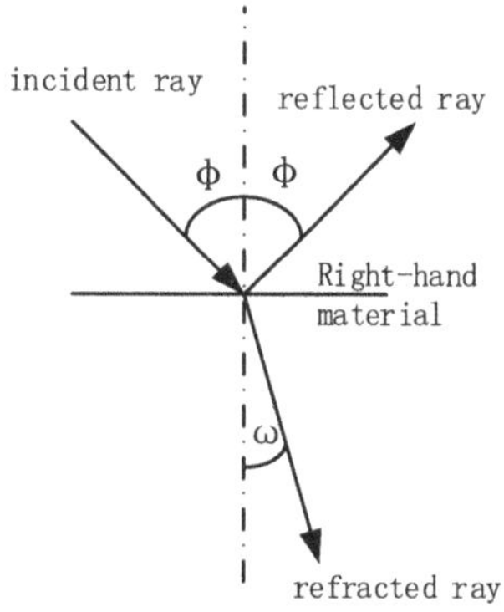

Fig. 5.14 Relation of reflection and reflection of right-handed material

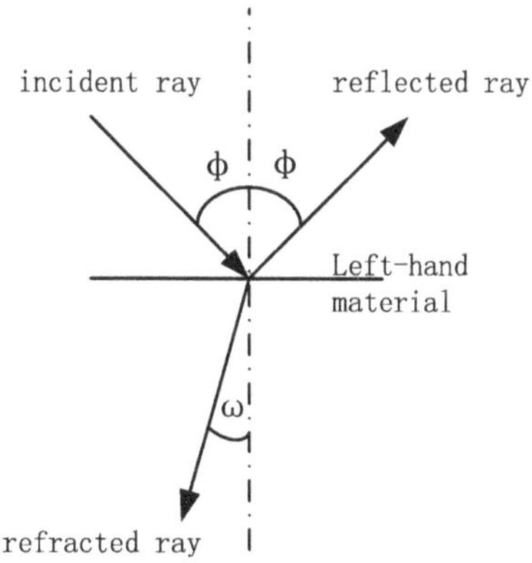

Fig. 5.15 Relation of reflection and reflection of left-handed material

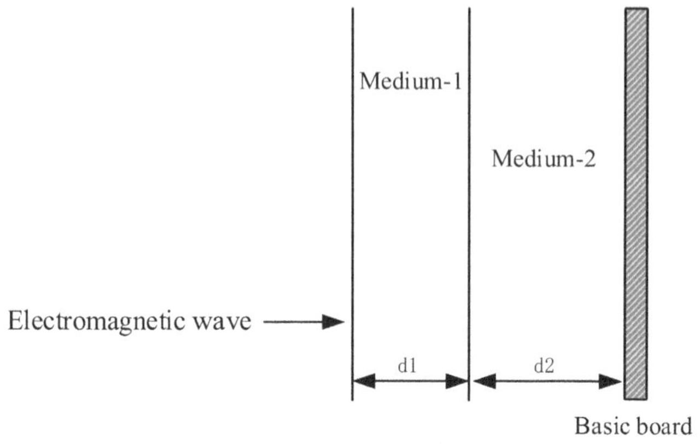

Fig. 5.16 Double layers absorbing material construction

When both Medium-1 and Medium-2 are right-hand materials, they are called right-right hand materials, which is the most common double-layer absorbing material structure. When medium-1 is a left-hand material and medium-2 is a right-hand material, it is called a left-right hand material.

Absorbing material property of the material is expressed by reflection loss R, which is defined as

$$R = -20\lg|r|\ [\mathrm{dB}] \tag{5.1}$$

In formula (5.1), the larger is R value; the larger the R value, the better the absorbing wave character of the material. $|r|$ is reflection coefficient which is depended on the electromagnetic parameter of different layer medium and expressed in the following:

$$|r| = \left| r(\varepsilon_i^{'}, \varepsilon_i^{''}, \mu_i^{'}, \mu_i^{''}, d_i) \right| \tag{5.2}$$

Here, the real and imaginary parts of the dielectric constant ε_i of the i-th layer material are denoted by $\varepsilon_i^{'}$ and $\varepsilon_i^{''}$, respectively, and the real and imaginary parts of the magnetic permeability μ_i are denoted by $\mu_i^{'}$ and $\mu_i^{''}$, respectively, and d_i is the thickness of i-th layer material.

The experimental results show that the absorbing characteristics of the left-right hand absorbing structure are significantly better than those of the conventional right-right hand absorbing structure in the 0.4 GHz to 10 GHz band (Table 5.4).

The absorbing wave feature curve of the above two kinds of double layer absorbing structure is shown in Fig. 5.17 and Fig. 5.18, respectively. In the figure, $R_{\max}$ is absorbing peak value, R_ is average reflection loss, and Δf is the frequency range of more than 20 dB reflection loss, which is called as the absorbing wave bandwidth, and the unit is GHz. $f_{\max}$ is peak frequency.

Table 5.4 Ranges of parameters and optimization of the two types of absorbing structures

Absorbing structure	Medium sequence	ε'	ε''	μ'	μ''	d/mm
left-right hand	LHM	−1~−50	0~0.00001	−1~−50	0~0.00001	1
	RHM	1~50	0.001~0.6	1~50	0.1~0.6	3
right-right hand	RHM	1~50	0~1	1~50	0~1	1
	RHM	1~50	0.001~0.6	1~50	0.1~0.6	3

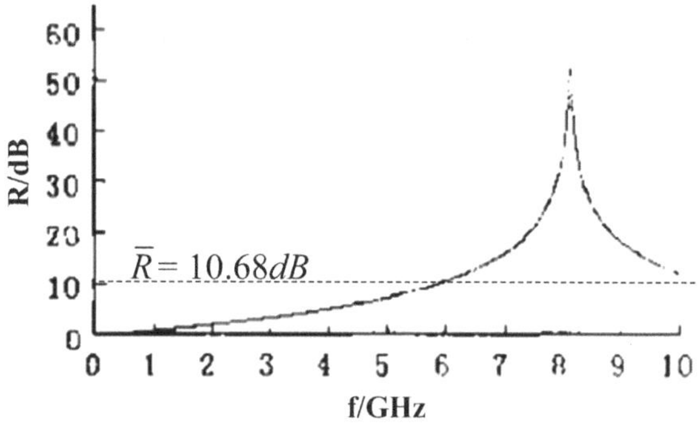

Fig. 5.17 Absorbing wave character curve of double layers absorbing structure of right–right-handed material

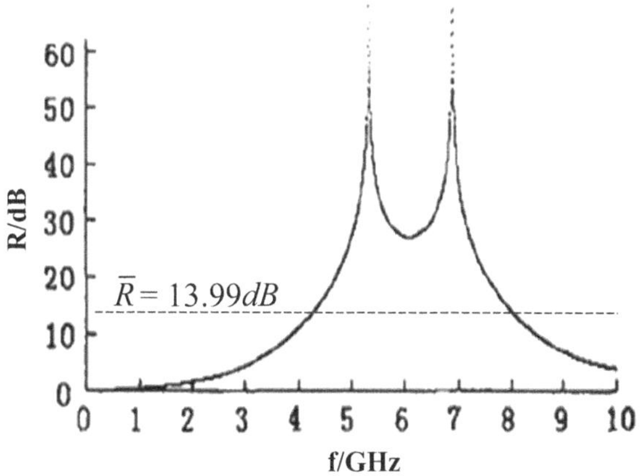

Fig. 5.18 Absorbing wave character curve of double layers absorbing structure of left–right-handed material

There is solo absorbing peak shown in Fig. 5.17 "absorbing wave character curve of double layers absorbing structure of right–right-handed material", in which the absorbing peak value R_{max} is about 55 dB and the average reflection loss R_ is 10.68 dB. The absorbing wave bandwidth is about 2 GHz when the reflection loss is 20 dB.

There are two absorbing peaks shown in Fig. 5.18 "absorbing wave character curve of double layers absorbing structure of left–right-handed material", in which the absorbing peak value R_{max} is more than 60 dB and the average reflection loss R_ is 13.99 dB. The absorbing wave bandwidth is obviously changing larger, when the reflection loss is 20 dB. The result is indicated that the introduction of left-handed material is the benefit of broadening the bandwidth of absorbing body and improving the absorbing peak to realize good matching of the material general electromagnetic parameter.

In conclusion, the introduction of double layers absorbing structure of left-handed material improves the absorbing performance as a whole, which includes peak value, bandwidth, and low- and high-frequency band absorptivity; the absorbing wave performance is improved obviously. The electromagnetic information leakage protection technology will be further developed with the deeper research of the absorbing wave feature of left-handed material.

5.3.2 Carbon Fiber Composite Material

With the development of modern electromagnetic information technology, conductive polymer shell is more and more widely popular due to the fine electromagnetic feature and the advantages of lower density and higher strength and so on. Gradually, it replaces the former metal shell of information facilities and widely applies in computer and electronic devices.

Carbon fiber of chemical nickel plating forms a composite coating on the surface after the chemical nickel plating which is a good combination between plating layer and carbon fiber. Carbon fiber of T-300 is made in TORAY of Japan, the original diameter of carbon fiber is 7 μm, and the nickel layer is uniform and compact by the technology of chemical nickel plating (the thickness is about 1 μm), as shown in Fig. 5.19.

Carbon fiber-plated nickel/epoxy resin composite material is compressed after the epoxy resin is mixed with curing agent in a certain percentage and added into a certain quantity of carbon fiber nickel plating. Many scattering interfaces of resin-nickel fiber are formed in the compression process of the composite material, which can be formed only one time of the electromagnetic wave reflection on the surface of the composite material. However, several times the reflection between the interfaces

Fig. 5.19 Sketch map of carbon fiber of chemical nickel plating

is generated in the inner material, so the total shielding effectiveness of the composite material is increased due to the reduction of strength of electromagnetic wave in great degree. Meanwhile, nickel coated fiber shows the layered network structure itself and also generates the induced current to decrease the electromagnetic energy. When the packing density of fiber coated nickel is increased and the electromagnetic energy is also largely reduced, meanwhile, the frequency range of electromagnetic wave is widely shielded.

The resistance is rapidly declined compared with the carbon fiber precursor after the surface of carbon fiber is chemically coated; the decreasing resistance value improves the conductivity of the composite material, and finally, the shielding effectiveness of the composite material is also strengthened. Simultaneously, the higher is the additional volume of carbon fiber nickel plating, the lower is the electrical resistivity of the composite material, the higher is the electromagnetic shielding capacity also. But attention should be paid that the shielding effectiveness of the electromagnetic wave isn't improved with the rise of frequency for the carbon fiber composite material nickel plating. For the lower frequency band of the electromagnetic wave (kHz frequency band), the shielding capacity of the composite material is mainly influenced by material characteristic parameter, even though the content of carbon fiber nickel plating in the composite material is changed, and total shielding capacity can't be largely changed. For the mid-high-frequency band of the electromagnetic wave (GHz frequency band), the shielding effectiveness of the composite material mainly depends on the electrical resistivity of the composite material; the smaller the resistivity of the material, the better the shielding effectiveness of the composite material.

Chapter 6
Electromagnetic Information Leakage and Protection

Computer as information processing facility is pervasive in modern society, and information processing includes calculation, storage, transmission, exchange, and so on. Components of computer play important roles during such processes; in the meantime, electromagnetic security issue exists. So, analysis of leakage sources and corresponding characteristics, and finding their effective solutions become main research issue. According to Chap. 4, a three-layered protective strategy was proposed. In this chapter, by analyzing specific cases of electromagnetic information leakage and protection, it was further verified that minimizing or keeping constant change of electromagnetic radiation signal is one of the most effective means.

6.1 Electromagnetic Information Security of Computer System

Computer is a complete information process system through coordination of different parts responsible for different functions. As each component of computer is different in operating principle and function, characteristics of electromagnetic information leakage are also different, thus relevant protection strategies would not be the same. Therefore, it is necessary to understand computer's main composition and structure and to analyze the characteristics of electromagnetic information leakage of each component.

6.1.1 Computer System Structure

Simply put, computer system constitutes hardware and software, which have layered structure, as shown in Fig. 6.1. The layers are divided by function; however, division of layers is not fixed and is changing as well.

T. Liu and Y. Li, *Electromagnetic Information Leakage and Countermeasure Technique*,
https://doi.org/10.1007/978-981-10-4352-9_6

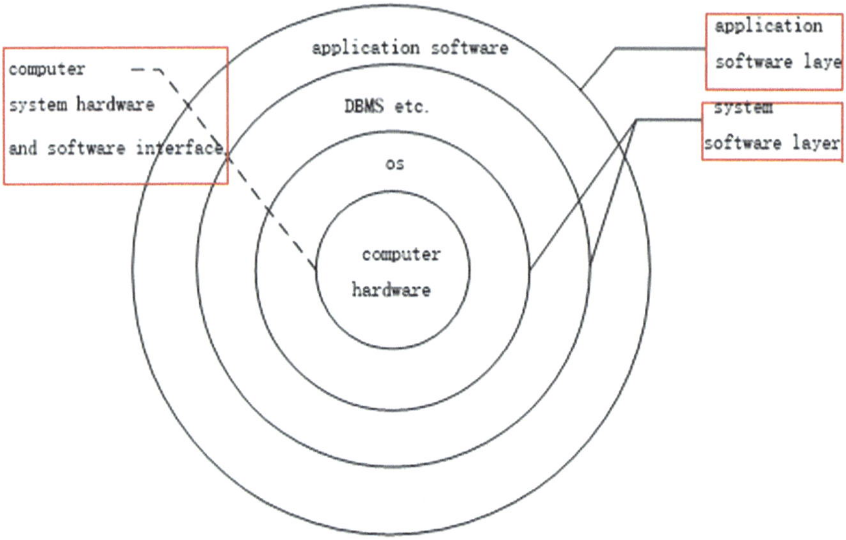

Fig. 6.1 Computer system structure

Computer hardware is composed of various components, which can be divided into host and its peripherals. Host is the main part of a computer system, whose core components include memories and CPU, and they are integrated on motherboard. Computer software is classified into system software and application software, which construct computer software system in a layered fashion.

Under normal circumstances, some parts of computer system are necessary and critical, and thus forms the relative stable and common parts in computer system [53], as shown in Fig. 6.2. The composition of the computer system would increase or decrease according to specific applications.

As shown in Fig. 6.2, electronic components of computer hardware system are the most primary electromagnetic information leakage source, and relation between the hardware system composition and each component is shown in Fig. 6.3.

6.1.2 Electromagnetic Information Protection Strategy of Computer System

Features of the electromagnetic information leakage are different in each component of computer hardware system, accordingly their protection measures also differ, but the applied protection principle is the same. According to the fourth chapter's three-layer protection strategy, the primary protection measure for electromagnetic information leakage starts from computer shielding method, that is, the first layer "leakage resistance", whose purpose is to decrease electromagnetic information leak-

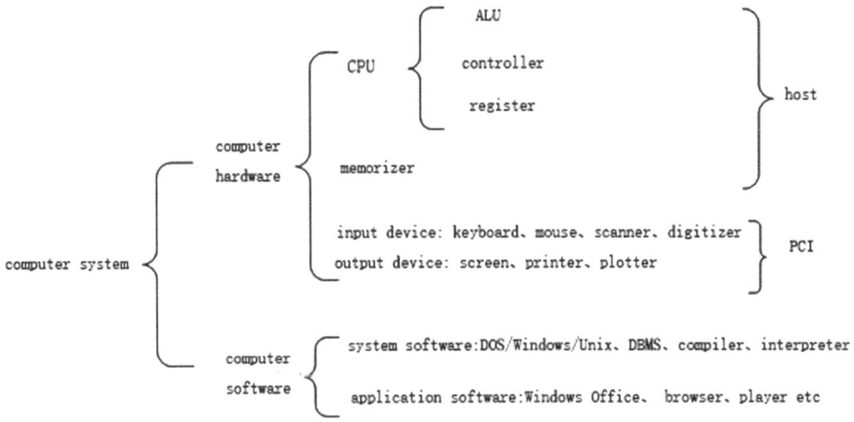

Fig. 6.2 Common configuration of computer system

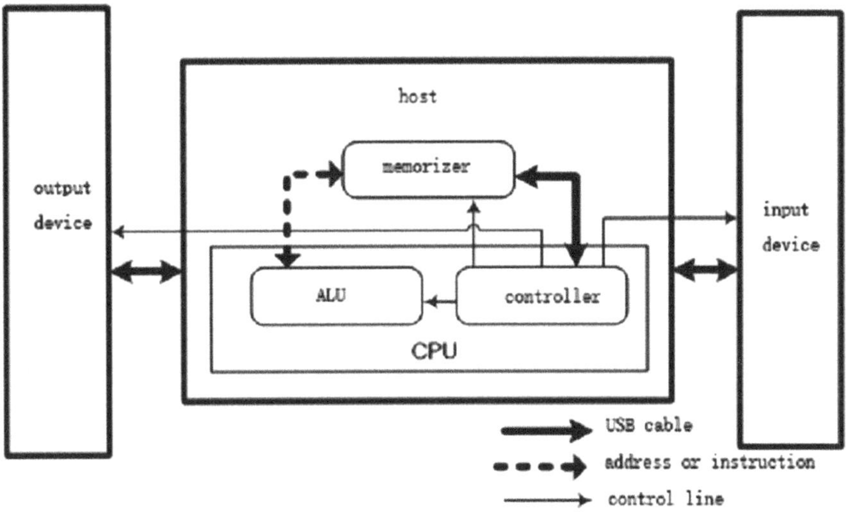

Fig. 6.3 Components construction of computer hardware system

age by methods as isolation, wave filtering, shielding, and so on. Because the above shielding method can't block the electromagnetic information leakage completely, and it is necessary to further incorporate the second layer "interception resistance" measure, which apply means of minimum change or constant change of electromagnetic radiation signals to increase interception difficulty. Even though the above two steps are adopted, it is still possible for electromagnetic information leakage, and thus the third layer "reproduction resistance" is added for anti-reconstruction attempt through intercepted electromagnetic information, for example, the use of coding technique can mutate electromagnetic pattern so as to make the reproduction of original information impossible.

Although software does not generate electromagnetic radiation directly, protection measures used in software layer such as software TEMPEST font technology can be necessary and effective. In fact, software measures have the advantages of lower cost, flexible to change, easy to implement, and imperceptibility for protection of electromagnetic information leakage compared to hardware protection measures. Therefore, protection of electromagnetic information leakage from software perspective is an important way and has better prospect for development.

In reality, there is no absolute information security. Depending on specific electromagnetic information leakage problem, considering all factors and comprehensive use of "leakage resistance", "interception resistance", and "reproduction resistance" strategies would be a practical solution for better protection of sensitive information.

6.2 Electromagnetic Information Leakage Source Features of Computer System

In computer system, components and their connection generate various electromagnetic radiations. In Fig. 6.3, different components have different degrees of leakage. Furthermore, as mechanism and function of components are different, different electromagnetic leakage characteristics will occur [18, 20, 22, 26, 54–59].

The electromagnetic leakage characters of computer need to be tested in special environment, and test facilities normally include antenna, primary amplifier, measurement receiver, and data recording devices shown in Fig. 6.4. The electromagnetic information leakage characters of main units of computer are illuminated by the following test data.

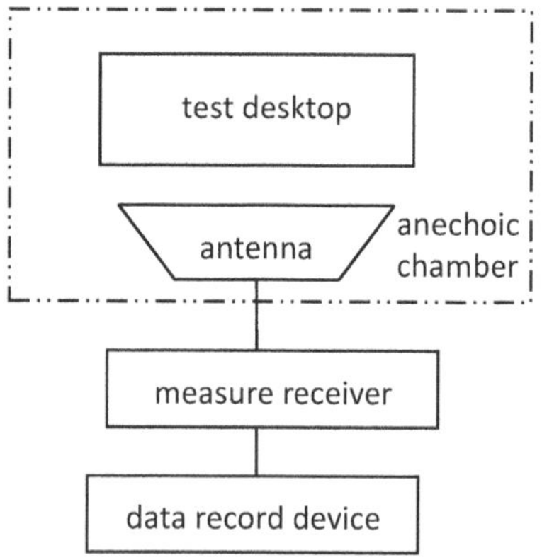

Fig. 6.4 Simple test environment sketch

6.2.1 Electromagnetic Leakage Characters of Displayers

Tested computer is placed on the table, and the sweep frequency test is operated in 1 m between the test antenna and the tested computer. The characters of the electromagnetic radiation frequency spectrum are shown in Figs. 6.5 and 6.7 in the condition of normal operation of the tested computer and the frequency range is between 10 kHz and 18 GHz.

The No. 1 device set under testing, denoted as DUT-I, includes computer host, LCD, keyboard, and mouse. The frequency spectrum of electromagnetic radiation is shown in Fig. 6.5.

The No. 2 device set, denoted as DUT-II, includes computer host, CRT display, keyboard, and mouse. The frequency spectrum of electromagnetic radiation is shown in Fig. 6.6.

The No. 3 device set, denoted as DUT-III, only includes computer host without peripherals, such as display, keyboard, mouse, etc. The frequency spectrum of electromagnetic radiation is shown in Fig. 6.7.

The characters of electromagnetic radiation are shown in the following when the computer is operated.

(1) Figure 6.5 shows a computer system with an LCD display. As can be seen from the figure, the impact of the LCD display on the whole machine is concentrated

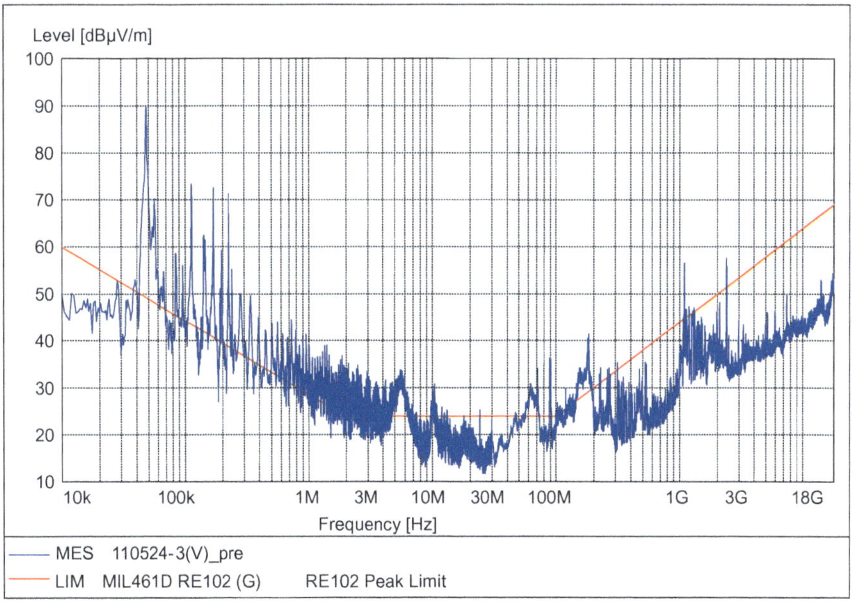

Fig. 6.5 Electromagnetic radiation characters of DUT-I (configuration: host, LCD, keyboard, and mouse)

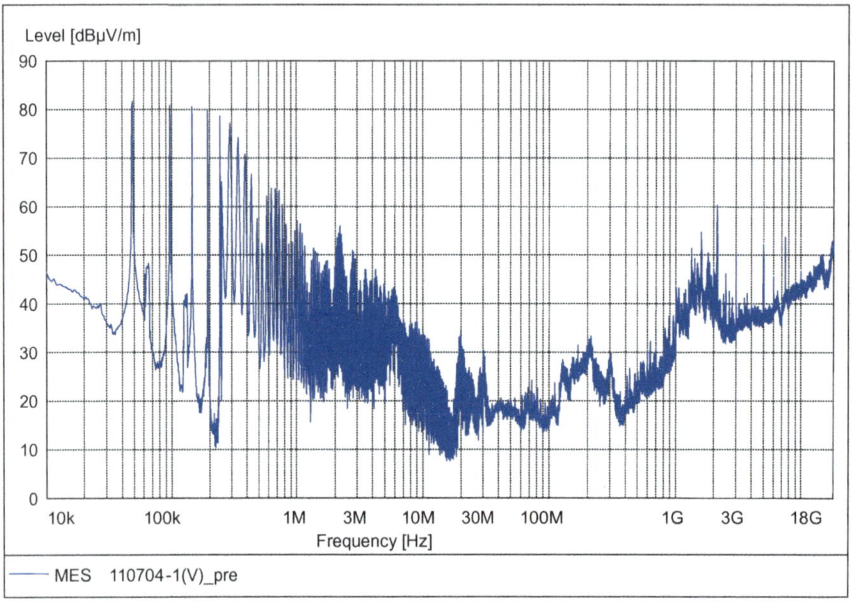

Fig. 6.6 Electromagnetic radiation characters of DUT-II (configuration: host, CRT, keyboard, and mouse)

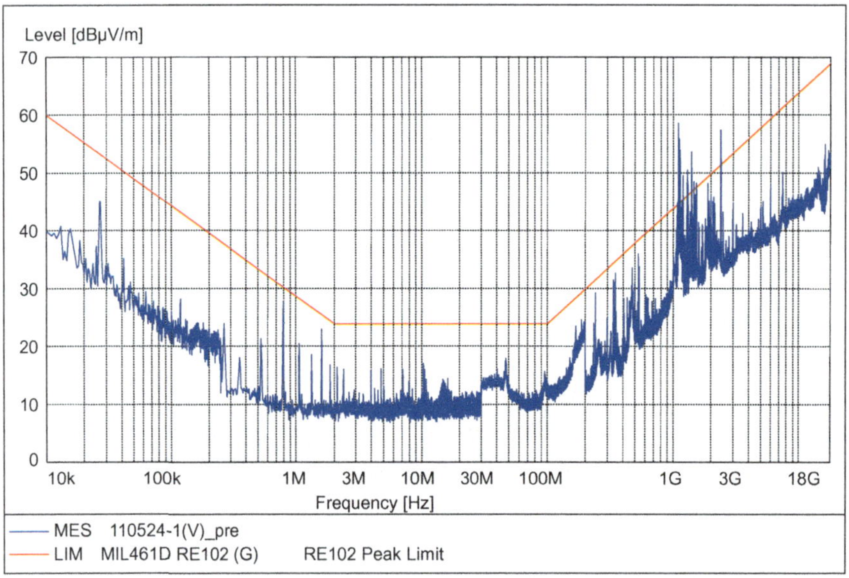

Fig. 6.7 Electromagnetic radiation characters of DUT-III (configuration: host)

in the range of 10 kHz–30 MHz, where the radiation intensity at 50 kHz is the highest with the amplitude of 90 dBμV/m, and the maximum radiation intensity between 110 and 220 kHz is about 74 dBμV/m. The radiation intensity between 300 kHz and 10 MHz shows a linear decreasing trend. The radiation intensity is attenuated from 50 to 25 dBμV/m; the radiation intensity between 10 and 30 MHz is relatively stable; the radiation intensity between 30 MHz and 18 GHz shows a linear growth. The radiation intensity value increases from 20 to 55 dBμV/m, and there are two separate peaks at 1.1 and 2.4 GHz, with the radiation intensity of 57 dBμV/m.

(2) Figure 6.6 shows a computer system with a CRT liquid crystal display. It can be seen from the figure that the impact of the CRT monitor on the whole machine is concentrated in the range of 10 kHz–30 MHz, where the radiation intensity at 50–160 kHz is the highest with the amplitude of 82 dBμV/m. The radiation intensity between 200 kHz and 30 MHz shows a linear decreasing trend. The radiation intensity is attenuated from 80 to 30 dBμV/m; the radiation intensity between 30 and 100 MHz is relatively stable, and the radiation intensity value is approximately 20 dBμV/m; the radiation intensity between 100 MHz and 18 GHz shows a linear growth trend. The radiation intensity value increases from 20 to 55 dBμV/m. There is a separate peak at 2.1 GHz with the radiation intensity of 61 dBμV/m.

(3) Figure 6.7 shows the radiation characteristics of a computer system that does not have peripherals such as monitors, keyboards, and mouse. It can be seen from the figure that the radiation intensity between 10 and 500 kHz shows a linear decreasing trend, the maximum radiation intensity is attenuated from 40 to 15 dBμV/m, with a separate peak at 28 kHz whose radiation intensity is about 46 dBμV/m. The radiation intensity between 500 kHz and 30 MHz is relatively stable, and the radiation intensity value is maintained at about 15 dBμV/m, among which peaks appear at 510 kHz, 800 kHz, 1.1 MHz, and 1.7 MHz, and the maximum amplitude is about 27 dBμV/m. The radiation intensity between 30 and 100 MHz is attenuated from 15 to 12 dBμV/m. The radiation intensity between 100 MHz and 18 GHz shows a linear growth trend, radiation intensity values increases from 15 to 55 dBμV/m, and there are two separate peaks located at 1.1 and 2.4 GHz with radiation intensity of 57 dBμV/m.

(4) By comparison of the above Figs. 6.5, 6.6, and 6.7, we can conclude that among frequency spectrum of computer system, display panel mainly lies in the lower part, but has higher radiation intensities. Obviously, CRT and LCD displays have different radiation intensities. CRT has higher intensities than LCD except at point of 50 kHz.

(5) The above is the spectrum test data of the display. The following are the time-domain test data of the liquid crystal display and the CRT display. The test environment is as same as that of the above, the test equipment is digital oscilloscope and loop antenna, the distance between the loop antenna and display is 30 cm. The time-domain radiation characters are tested between LCD display and CRT display, respectively, in three kinds of status, which are display characters, white screen, and blank screen, and the test results are shown in Fig. 6.8.

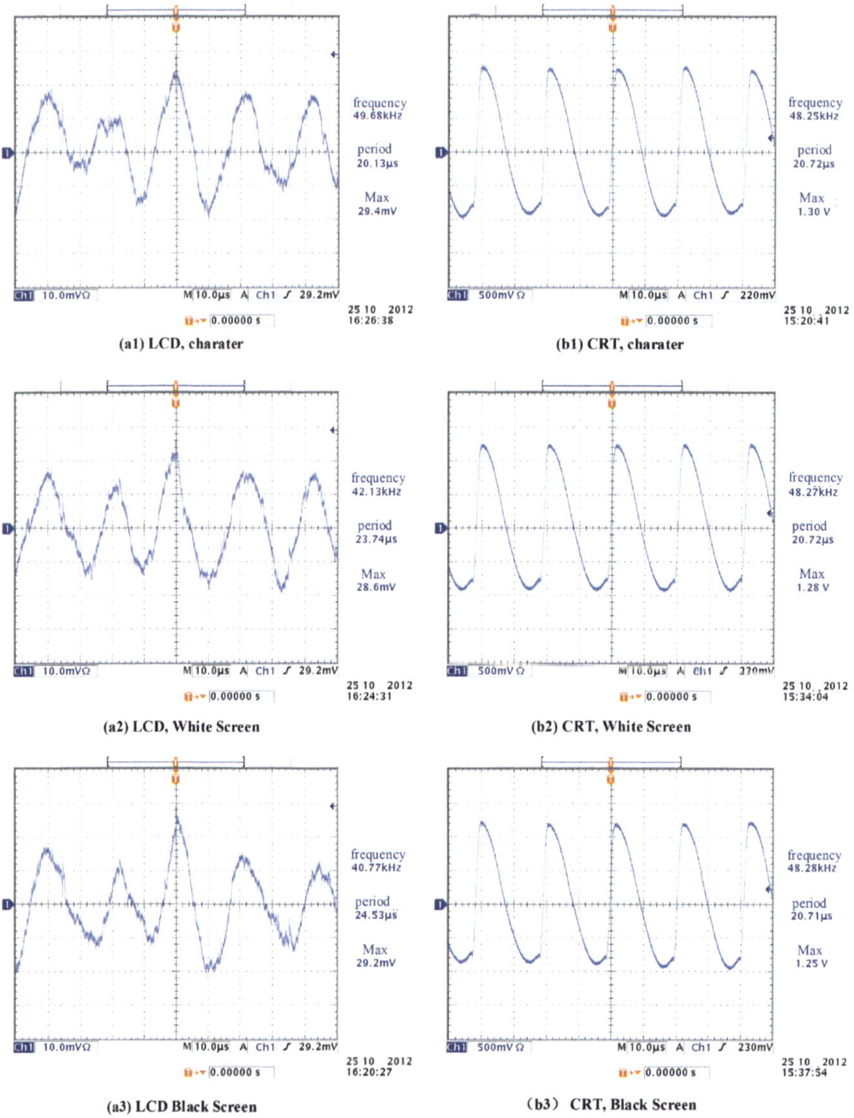

Fig. 6.8 Time domain diagram of three displays

The oscilloscope's parameter selection must be able to match the waveform change, because the radiation value of the LCD monitor is less than the radiation value of the CRT monitor, and the test parameter settings are not exactly the same.

The time-domain figure of LCD display is shown in Fig. 6.8a, in which the parameter of oscilloscope is 10 mV and 10 μs and the time-domain figure of CRT display is shown in Fig. 6.8b, in which the parameter of oscilloscope is 500 mV and 10 μs.

The electromagnetic radiation of the display is shown in the following characters from the tested data:

(1) Figure 6.8a shows radiation waveforms of LCD in three scenes: displaying characters, white screen, and black screen. The three waveforms have slight deviations, especially at the positions of peaks and troughs. When displaying characters, maximum wave peak amplitude is 29.4 mV, average wave peak is 19.28 mV, maximum peak–trough amplitude is 44 mV, minimum peak–trough amplitude is 15 mV, and the period is 20.13 μs. When white screen is shown, maximum wave peak amplitude is 28.6 mV, average wave peak is 19.52 mV, maximum peak–trough amplitude is 42 mV and minimum peak–trough amplitude is 30 mV, the period is 23.74 μs. When black screen is shown, maximum wave peak amplitude is 29.2 mV, average wave peak amplitude is 18.24 mV, maximum peak–trough amplitude is 48 mV, minimum peak–trough amplitude is 26 mV, and the period is 24.53 μs.

(2) Figure 6.8b shows radiation waveforms of CRT in three scenes: displaying characters, white screen, and black screen. The three waveforms are basic consistent and smooth without distortion. When displaying characters, maximum wave peak amplitude is 1.3 V, average wave peak amplitude is 1.272 V, maximum peak–trough amplitude is 2.3 V, minimum peak–trough amplitude is 2.15 V, and the period is 20.72 μs. When white screen is shown, maximum wave peak amplitude is 1.28 V, average wave peak is 1.276 V, maximum peak–trough amplitude is 2.18 V, minimum peak–trough amplitude is 2.175 V, and the period is 20.72 μs. When black screen is shown, maximum wave peak amplitude is 1.25 V, average wave peak amplitude is 1.23 V, maximum peak–trough amplitude is 2.25 V, minimum peak–trough amplitude is 2.11 V, and the period is 20.71 μs.

(3) The oscilloscope amplitude ranges in Fig. 6.8b is 50 times smaller than in Fig. 6.8a. The reason is that the radiation strength of LCD display is far less than that of CRT display. There is a difference of the imaging mechanism between CRT display and LCD display, and there isn't circuit including the high-pressure deflection coil and amplifier and or so in LCD display which is necessary for CRT display, besides the operating voltage of LCD is less than that of CRT display. So, the radiation value of LCD display is lower than that of CRT display.

6.2.2 CPU Electromagnetic Leakage Characters

The following is a spectrum scan of different CPUs, as shown in Fig. 6.9, analyzing electromagnetic radiation leakage characteristics of CPU through spectrogram. The detailed configuration of tested computer is listed in Table 6.1. The electromagnetic radiation leakage character of other component will be analyzed in the farther below:

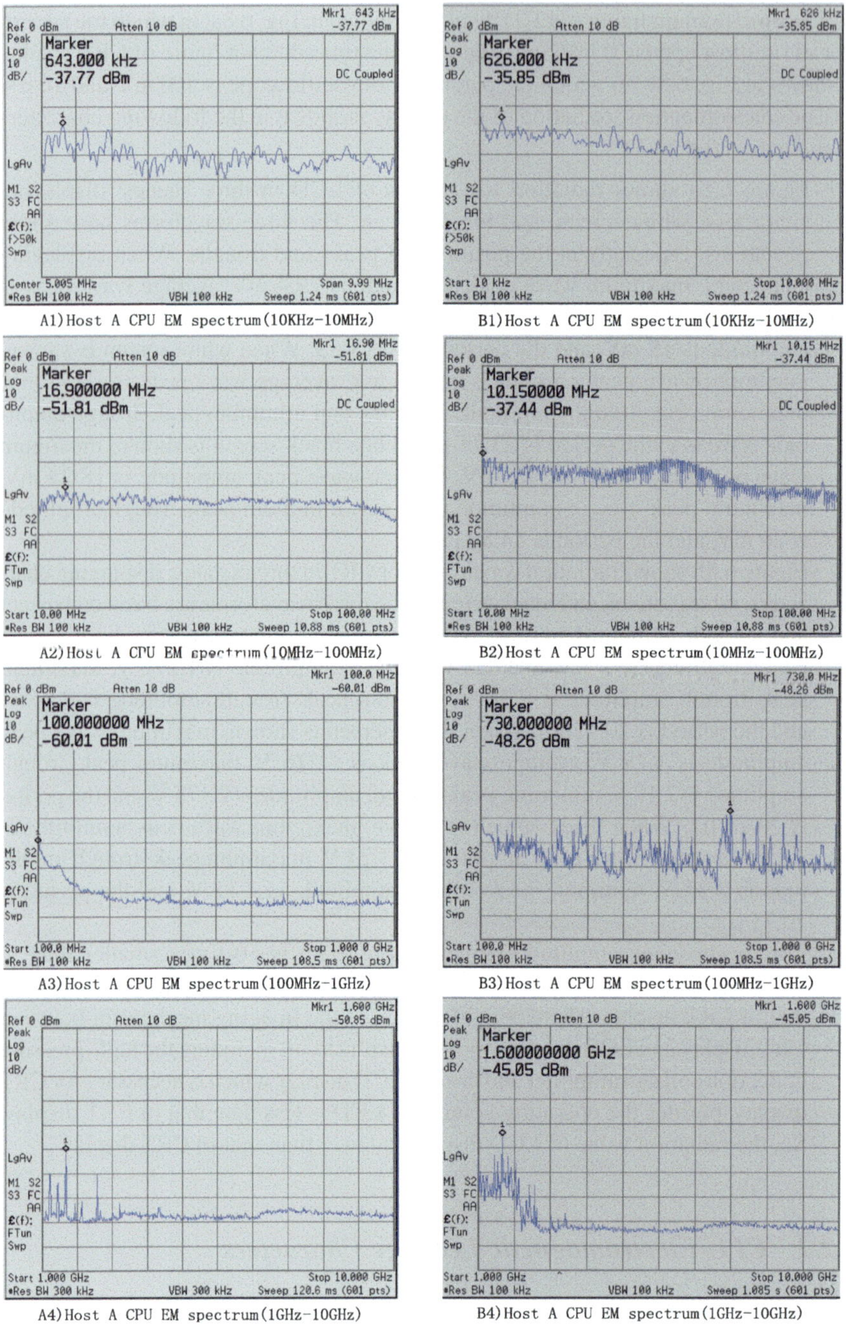

Fig. 6.9 CPU electromagnetic y spectrum

Table 6.1 Tested computer configuration

Modules	Computer system A	Computer system B
Power	Patriot	Jin Hetian
CPU	Intel E4500, 65 nm, clocked at 2.2 GHz, multiplier 11, FSB 200 MHz, front-side bus 800 MHz, supply voltage 1.3 V	Intel Pentium 4 630, 90 nm, clocked at 3.0 GHz, multiplier 15, FSB 200 MHz, front-side bus 800 MHz, supply voltage 1.25–1.40 V
Motherboard	MICRO-STAR	ASUS
Video card	NVDIA GeForce 9400	NVDIA Geforce4 MX440
Memory	DDR2 800	DDR2 667
Hard disk	Seagate, 160 GB, interface SATA/300, speed 7200	Maxtor, 80 GB, IDE interface, speed 5400

1. Electromagnetic Radiation of CPU in Host A

   (1) 10 kHz–10 MHz: maximum electromagnetic radiation intensity is −37.77 dBm (frequency point 653 kHz), radiation intensity is −60 to −37.7 dBm, and floating scope is 22.7 dBm.
   (2) 10–90 MHz: maximum electromagnetic radiation intensity is −51.81 dBm (frequency point 16.9 MHz), radiation intensity is −60 to −51.81 dBm, and floating scope is 8.19 dBm.
   (3) 90–280 MHz: the radiation intensity is decreased from −60 to −82 dBm, shown the linear decline trends.
   (4) 280 MHz–1 GHz: radiation intention is −82 to −80 dBm and floating scope is 2 dBm.
   (5) 1–3 GHz: maximum electromagnetic radiation intensity is −50.85 dBm (frequency point 1.6 GHz), radiation intensity is −80 to −50.85 dBm, floating scope is 29.15 dBm, and amplitude of variation is larger.
   (6) 3–10 GHz: radiation intensity is −80 to −75 dBm and floating scope is 5 dBm.

2. Electromagnetic Radiation of CPU in Host B

   (1) 10 kHz–10 MHz: maximum electromagnetic radiation intensity is −35.85 dBm (frequency point 626 kHz), radiation intensity is −58 to −35.85 dBm, and floating scope is 22.15 dBm.
   (2) 10–90 MHz: maximum electromagnetic radiation intensity is −37.44 dBm (frequency point 10.15 MHz), radiation intensity is −60 to −37.44 dBm, and floating scope is 22.56 dBm.
   (3) 100 MHz–1 GHz: maximum electromagnetic radiation intensity is −48.26 dBm (frequency point 730 MHz), radiation intensity is −80 to −48.26 dBm, and floating scope is 31.74 dBm.
   (4) 1–4 GHz: maximum electromagnetic radiation intensity is −45.05 dBm (frequency point 1.6 GHz), radiation intensity is −83 to −45.05 dBm, floating scope is 37.95 dBm, and amplitude of variation is larger.

(5) 3–10 GHz: radiation intensity is −83 to −80 dBm and floating scope is 3 dBm.

Through comparison, it is found that the intensity of electromagnetic radiation in the vicinity of the dominant frequency changes the most. Therefore, this active region of the radiation zone is further subdivided and scanned, as shown in Fig. 6.10.

(1) Radiation of CPU in host A. In 1–2 GHz frequency band, maximum electromagnetic radiation intensity is −41.95 dBm (frequency point is 1.607 GHz), radiation intensity is −73 to −41.95 dBm, and floating scope is 31.05 dBm, and there are many spikes in this band. In 2–3 GHz frequency band, maximum elec-

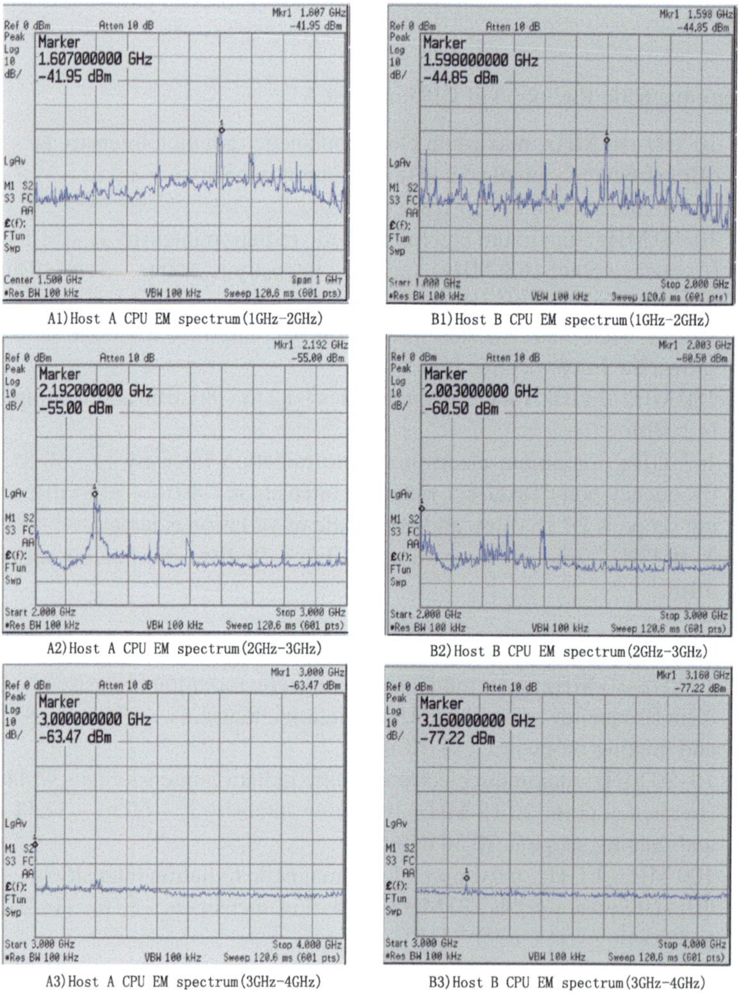

Fig. 6.10 CPU electromagnetic y spectrum 2

tromagnetic radiation intensity is −55.00 dBm (frequency point is 2.192 GHz), radiation intensity is −84 to −55.00 dBm, and floating scope is 31.05 dBm, and there are fewer spikes in this band (in which the radiation intensity is basic stability near −82 dBm, after 2.52 GHz). In 3–4 GHz frequency range, the radiation intensity is basic stability besides this point, intensity is −83 to −77 dBm and floating range is 6 dBm.

(2) Radiation of CPU in host B. In 1–2 GHz frequency range, maximum electromagnetic radiation intensity is −44.85 dBm (frequency point is 1.598 GHz), radiation intensity is −80 to −44.85 dBm, and floating scope is 31.15 dBm, and there are more spikes in this band. In 2–3 GHz frequency range, maximum electromagnetic radiation intensity is −60.50 dBm (frequency point is 2.003 GHz), radiation intensity is −84 to −60.50 dBm, and floating scope is 23.50 dBm, and there are less spikes in this band (in which the radiation intensity is basic stability near −82 dBm, after 2.45 GHz). In 3–4 GHz frequency range, other radiation intensity is basic stability besides this point, intensity is −83 to −80 dBm and floating range is 3 dBm.

Electromagnetic leakage characteristics of the CPU can be summarized as follows:

The overall trend of radiation intensity is downward, radiation intensity of frequency band (10 kHz–1 GHz), before CPU frequency, is the larger than that of other bands. Electromagnetic radiation intensity varies mostly in the frequency band near the main frequency (1–4 GHz), and there are several radiation spikes in this band, especially with a maximum radiation intensity around 1.6 GHz. These radiation spikes are primarily formed by the CPU's main frequency, FSB, and front-side bus frequencies and their harmonics.

6.2.3 Electromagnetic Leakage Characters of Video Card

Spectrum scanning is operated for computer video card, illustrated in Fig. 6.11, electromagnetic leakage character of video card is analyzed by the frequency graphs.

1. Electromagnetic Radiation Character of Video Card for Host A

(1) 10 kHz–10 MHz: maximum electromagnetic radiation intensity is −29.72 dBm (frequency point is 1.825 MHz), radiation intensity is −70 to −29.72 dBm, and floating scope is 40.28 dBm. There is obviously periodic variation.

(2) 10–100 MHz: maximum electromagnetic radiation intensity is −30.49 dBm (frequency point is 20.65 MHz), radiation intensity is −60 to −30.49 dBm, and floating scope is 29.51 dBm. There is obviously periodic variation and the intensity changing range is shown downtrend.

(3) 100 MHz–1 GHz: maximum electromagnetic radiation intensity is −46.68 dBm (frequency point is 139.0 MHz), radiation intensity is shown the linear decline from 139.0 to 350 MHz and is maintained in −86 dBm approximately.

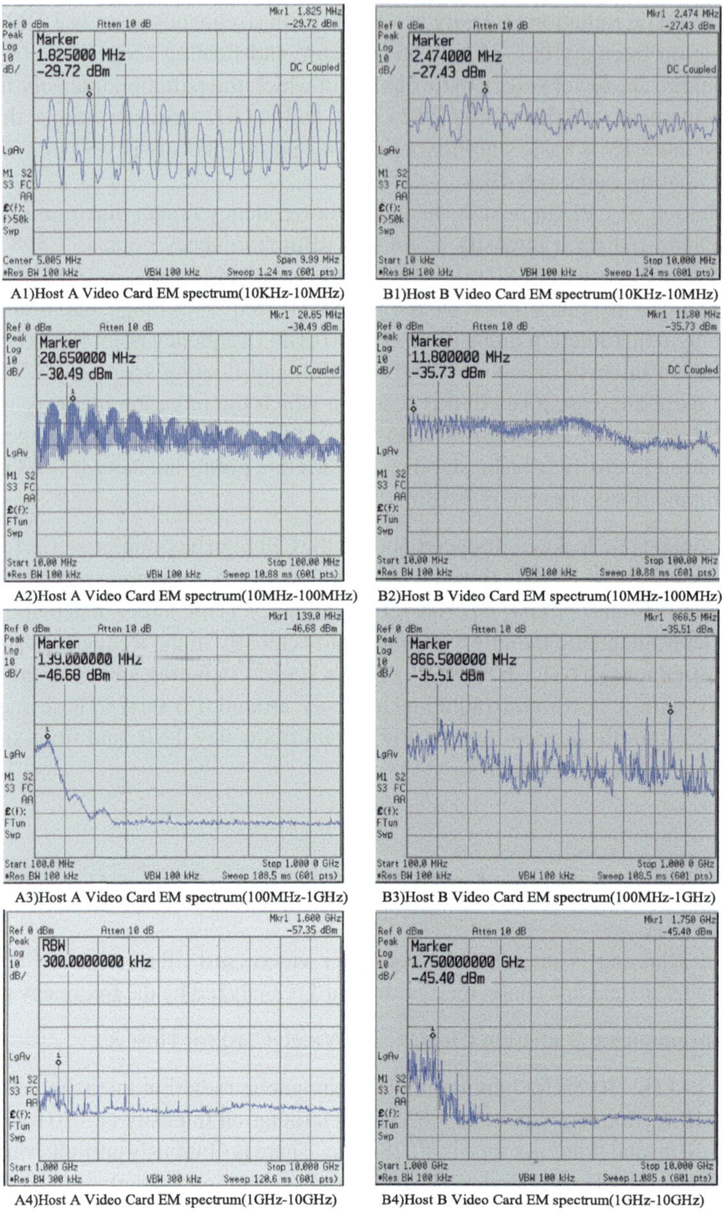

Fig. 6.11 Video card electromagnetic spectrums

(4) 1–10 GHz: maximum electromagnetic radiation intensity is −57.35 dBm (frequency point is 1.6 GHz), radiation intensity is −80 to −57.35 dBm, floating scope is 22.65 dBm, and there isn't obvious variation in 5.5 GHz.

2. Electromagnetic Radiation Character of Video Card for Host B

(1) 10 kHz–10 MHz: maximum electromagnetic radiation intensity is −27.43 dBm (frequency point is 2.47 MHz), radiation intensity is −50 to −27.43 dBm, and floating scope is 22.57 dBm. There is a periodic variation.
(2) 10–100 MHz: maximum electromagnetic radiation intensity is −35.73 dBm (frequency point is 11.80 MHz), radiation intensity is −55 to −35.73 dBm, and floating scope is 19.27 dBm, and the intensity is shown downtrend.
(3) 100 MHz–1 GHz: maximum electromagnetic radiation intensity is −35.51 dBm (frequency point is 866.50 MHz), radiation intensity is −72 to −35.51 dBm, and floating range is 26.49 dBm, and the intensity fluctuation is very active.
(4) 1–10 GHz: maximum electromagnetic radiation intensity is −45.40 dBm (frequency point is 1.75 GHz), radiation intensity is −83 to −45.40 dBm, and floating scope is 37.60 dBm, and there isn't obvious variation in 3.6 GHz.

After comparing and analyzing, the characters of electromagnetic leakage of video card are shown, with the increase of the frequency, electromagnetic radiation intensity presents downtrend, and there is special active frequency region. The active region of video card A is 10 kHz–350 MHz and the active region of video card B is 100 MHz–3.6 GHz, which keep the stabilization without signal radiation after certain high-frequency point.

6.2.4 Electromagnetic Leakage Characters of Hard Disk

The following is spectrum scan of computer hard disk, as shown in Fig. 6.12, analyzing electromagnetic radiation leakage characteristics of hard disk through the spectrogram:

1. Electromagnetic Radiation Character of Hard Disk for Host A

(1) 10 kHz–10 MHz: maximum electromagnetic radiation intensity is −48.17 dBm (frequency point is 1.009 MHz), radiation intensity ranges from −74 to −48.17 dBm, and floating scope is 25.83 dBm. There is clearly cyclic variation.
(2) 10–100 MHz: maximum electromagnetic radiation intensity is −62.51 dBm (frequency point is 80.05 MHz), radiation intensity ranges from −80 to −62.51 dBm, and floating scope is 17.49 dBm, and the intensity has shown slow uptrend.
(3) 100 MHz–1 GHz: maximum electromagnetic radiation intensity is −46.01 dBm (frequency point is 160.00 MHz), radiation intensity ranges from

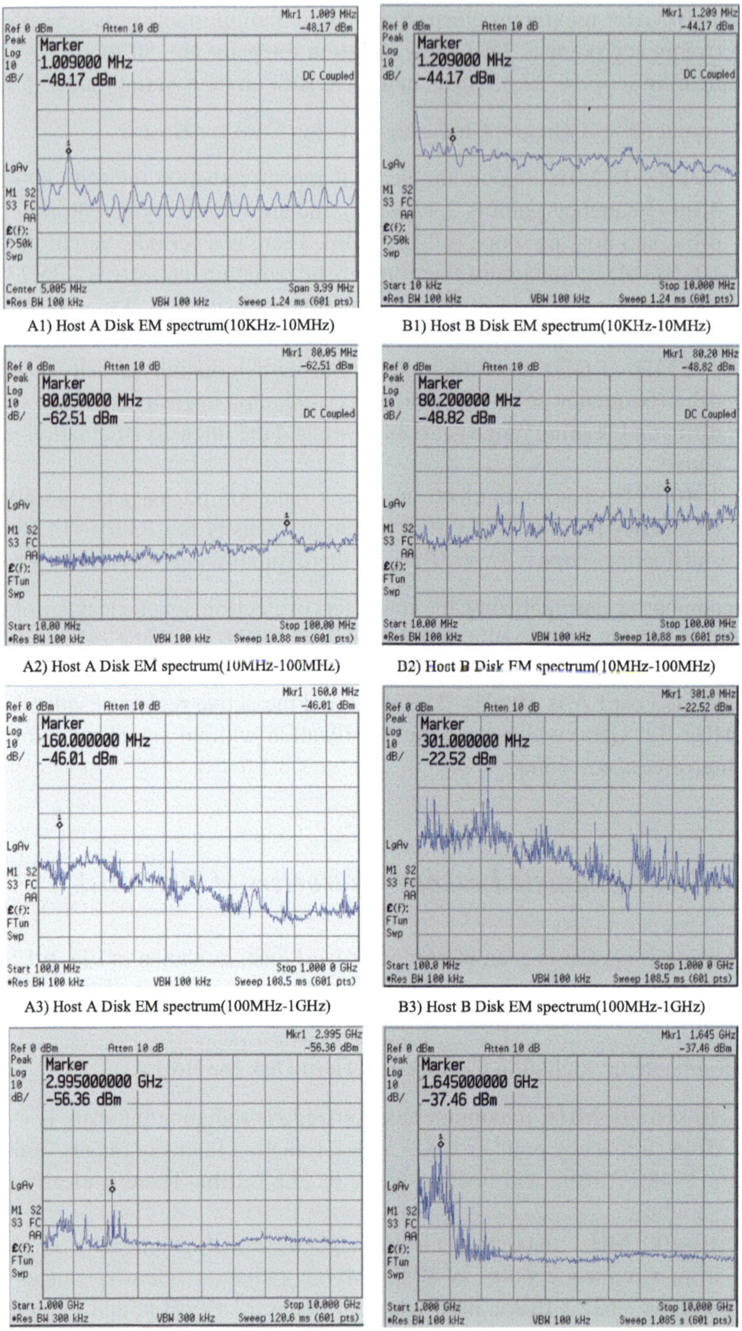

A1) Host A Disk EM spectrum(10KHz-10MHz)

B1) Host B Disk EM spectrum(10KHz-10MHz)

A2) Host A Disk EM spectrum(10MHz-100MHz)

B2) Host B Disk EM spectrum(10MHz-100MHz)

A3) Host A Disk EM spectrum(100MHz-1GHz)

B3) Host B Disk EM spectrum(100MHz-1GHz)

A4) Host A Disk EM spectrum(1GHz-10GHz)

B4) Host B Disk EM spectrum(1GHz-10GHz)

Fig. 6.12 Hard disk electromagnet spectrum

−84 to −46.01 dBm, and floating range is 37.99 dBm, and the intensity is very fluctuating.

(4) 1–10 GHz: maximum electromagnetic radiation intensity is −56.36 dBm (frequency point is 2.995 GHz), radiation intensity ranges from −80 to −56.36 dBm, and floating scope is 23.64 dBm, and there isn't obvious variation after 3.6 GHz.

2. Electromagnetic Radiation Character of Hard Disk for Host B

(1) 10 kHz–10 MHz: maximum electromagnetic radiation intensity is −32.0 dBm (frequency point is 10 kHz), radiation intensity ranges from −58 to −32.0 dBm, and floating scope is 26.0 dBm.

(2) 10–100 MHz: maximum electromagnetic radiation intensity is −48.82 dBm (frequency point is 80.20 MHz), radiation intensity is −72 to −48.82 dBm, and floating scope is 23.18 dBm, and the intensity shows slow uptrend.

(3) 100 MHz–1 GHz: maximum electromagnetic radiation intensity is −22.52 dBm (frequency point is 310.00 MHz), radiation intensity ranges from −80 to −22.52 dBm, and floating range is 57.48 dBm, and the intensity is very fluctuating.

(4) 1–10 GHz: maximum electromagnetic radiation intensity is −37.46 dBm (frequency point is 1.645 GHz), radiation intensity ranges from −83 to −37.46 dBm, and floating scope is 45.54 dBm, and there isn't obvious variation after 3.7 GHz.

Through the above analysis, it can be found that the electromagnetic leakage characteristics of hard disk are as follows: electromagnetic radiation intensity is high in the middle frequency band and low at both ends. Hard disks A and B have frequency bands with large radiation fluctuations. The frequency band of hard disk A is from 100 MHz to 3.6 GHz and the frequency band of hard disk B is from 100 MHz to 3.7 GHz.

6.2.5 Electromagnetic Leakage Characters of Power

Electromagnetic signal frequency spectrum of power is illustrated in Fig. 6.13.

1. Electromagnetic Radiation of Power A

(1) 10 kHz–100 MHz: maximum electromagnetic radiation intensity is −24.71 dBm (frequency point is 1.840 MHz), radiation intensity ranges from −51 to −24.71 dBm, and floating scope is 26.29 dBm.

(2) 100 MHz–1 GHz: maximum electromagnetic radiation intensity is −47.30 dBm (frequency point is 719.50 MHz), radiation intensity ranges from −80 to −47.30 dBm, and floating range is 32.70 dBm. The intensity fluctuation is quite operational.

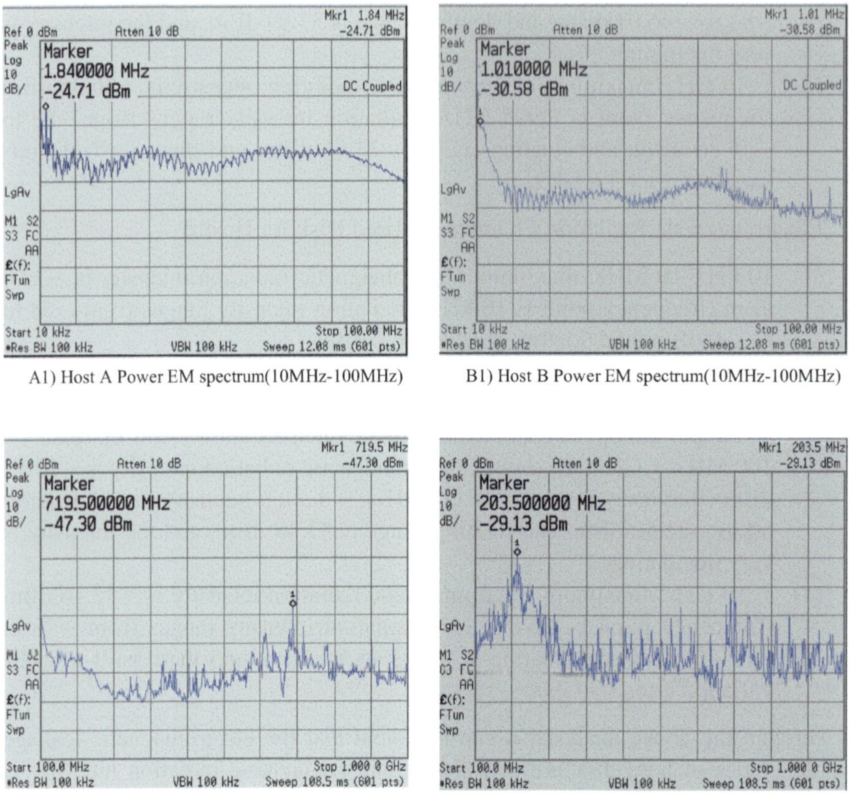

A1) Host A Power EM spectrum(10MHz-100MHz)

B1) Host B Power EM spectrum(10MHz-100MHz)

A2) Host A Power EM spectrum(100MHz-1GHz)

B2) Host B Power EM spectrum(100MHz-1GHz)

Fig. 6.13 Power electromagnetic spectrum

2. Electromagnetic Radiation of Power B

   (1) 10 kHz–100 MHz: maximum electromagnetic radiation intensity is −30.58 dBm (frequency point is 1.01 MHz), radiation intensity ranges from −65 to −30.58 dBm, and floating scope is 34.42 dBm.
   (2) 100 MHz–1 GHz: maximum electromagnetic radiation intensity is −29.13 dBm (frequency point is 203.50 MHz), radiation intensity ranges from −80 to −29.13 dBm, and floating range is 50.87 dBm. The intensity is very fluctuating.

By discussions mentioned above, it can be found that the electromagnetic leakage characteristics of the power supply are expressed as power supply has a significant fluctuation in frequency range of 100 MHz–1 GHz. The main reason is that power supply needs to provide a variety of operating voltages, and electromagnetic radiation of different step-down circuits and their cooling fans are spatially aliased, so the radiation intensity fluctuates greatly.

6.2.6 Electromagnetic Leakage Characters of Wireless Keyboard

Compared with wired keyboards, electromagnetic radiation of wireless keyboards is intentional, but there is information leakage. Each wireless keyboard has its own characteristics as given below:

(1) Different wireless keyboard is unique in working principle.
(2) Different wireless keyboard is different in signal frequency spectrum.

These features can be used to distinguish between different wireless keyboards. The following shows electromagnetic radiation of three wireless keyboard of different models in three different states: wireless connection but no button press, wireless connection and button press, and no wireless connection but button press, as shown in Fig. 6.14.

From Fig. 6.14, we can draw following conclusions:

(1) The working mechanisms of keyboard 1 and 2 are similar. After wireless connection is established, there will be no strong electromagnetic radiation if there is no button operation. However, Keyboard 3 is totally dissimilar. Once wireless connection is established, electromagnetic radiation is maintained, regardless of button operation.
(2) When the keyboard 1 and the keyboard 2 are pressed, electromagnetic radiation is generated, and the spectral characteristics are obvious. When the keyboard 3 is pressed, the difference in electromagnetic spectrum is not obvious to the former.
(3) When wireless connection is not established, keyboard 1 and the keyboard 2 will generate electromagnetic radiation when a button is pressed, but keyboard 3 does not emit electromagnetic waves under the same conditions.

So, it is possible to distinguish different wireless keyboards by analyzing spectrum and determine which keyboard generates the electromagnetic radiation.

6.3 Electromagnetic Information Leakage of Computer System and Typical Protection Cases

Electromagnetic information leakage cases of all parts of computer system are laid down in this chapter and protection strategy is pointed out.

6.3.1 Core Part—CPU

CPU is the core component of the computer and all data is processed by the CPU, and if an electromagnetic leakage occurs at the CPU, it is extremely harmful. Because

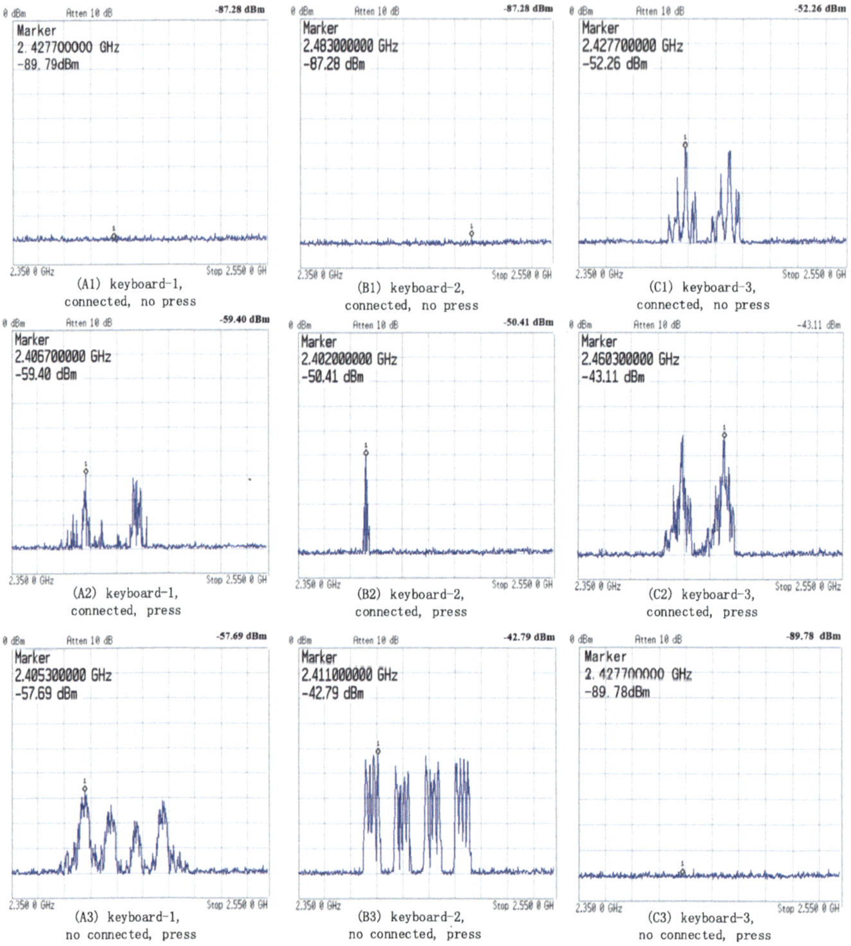

Fig. 6.14 Wireless keyboard electromagnetic spectrum

working voltage of CPU is not high, its electromagnetic radiation intensity is not strong. For protection strategy of CPU, "anti-leakage" is preferred and reinforcement shielding is the most effective method.

Actual measurement of electromagnetic radiation features of CPU has been already analyzed in detail in Sect. 6.2.2. The protection of CPU should be targeted, and its frequency band with strong electromagnetic radiation should be reinforced and shielded. In principle, selection of reinforcement shielding material should be based on main frequency of CPU.

6.3.2 *Output Device—Display*

As main component of computer system, display is the most important human–computer interaction interface. The information displayed is in line with human reading customs and has a clear meaning. At the same time, ciphertext cannot be read. So, it is difficult to protect data security by means of encryption. Once the information displayed in displays is intercepted and captured, the essential information leakage is easily caused. There are many types of displays, such as CRT monitors, LCD monitors, and LED monitors. It is the main source of leakage of computer electromagnetic radiation, according to actual measurement, no matter which kind of it.

The leakage of CRT monitor was discovered very early. As early as 1986, Van Eck intercepted and reproduced the information displayed on the CRT several meters away with simple equipment. LCD that appears after CRT is generally considered to be low-emission devices, but studies have shown that LCD also has electromagnetic information leakage, and the intercepted signals can be reproduced, which is no more difficult than intercepting of CRT. Although LCD has lower radiation, it has not yet met the requirements for electromagnetic information security.

The interception and reproduction of CRT display have been introduced in the previous chapters. Only the case of LCD is given here. Through the case, analyze the key leakage of LCD and give protection methods.

1. Interception and Reproduction Tests

The original image of LCD display of desk computer is shown in Fig. 6.15, reproduction image of intercepted and captured electromagnetic signal is shown in Fig. 6.16. The original image of LCD display of portable computer is shown in Fig. 6.17, reproduction image of intercepted and captured electromagnetic signal is shown in Fig. 6.18.

Fig. 6.15 Desktop LCD display image

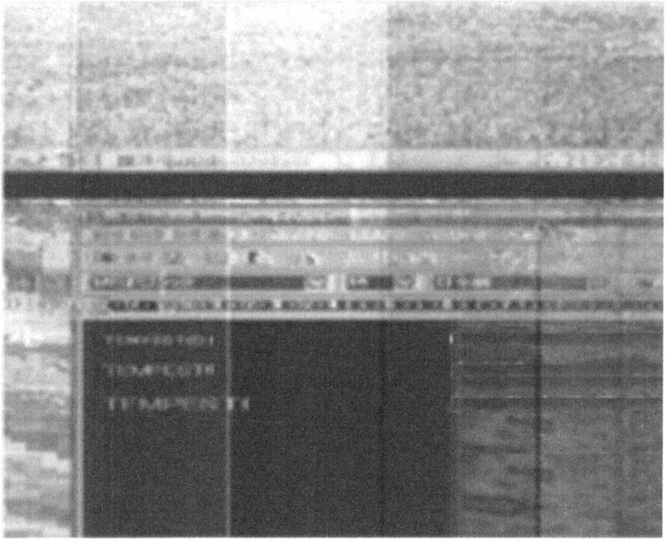

Fig. 6.16 Reproduced image of desktop LCD

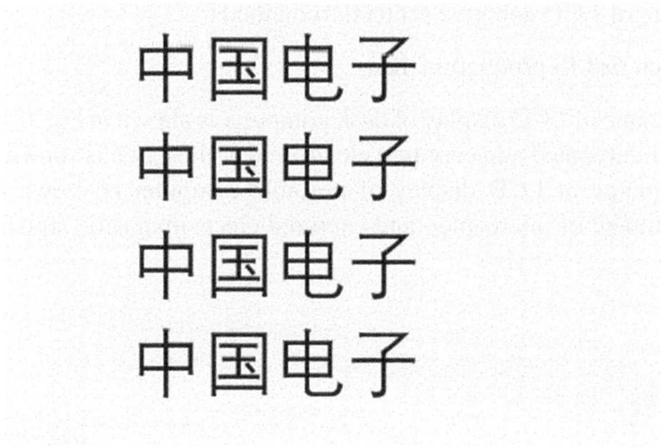

Fig. 6.17 Laptop LCD display image

2. Key Leakage Location Analysis

Main radiation locations of CRT display are high-speed electron beam and deflection components, main radiation locations of LCD display are panel key and transmission USB cables, and main radiation locations of display for portable computer are connection location of display and built-in data line beside of display, as shown in Fig. 6.19.

In summary, although various types of displays are different in construction and model, the main radiating parts of display are distributed around screen and its internal

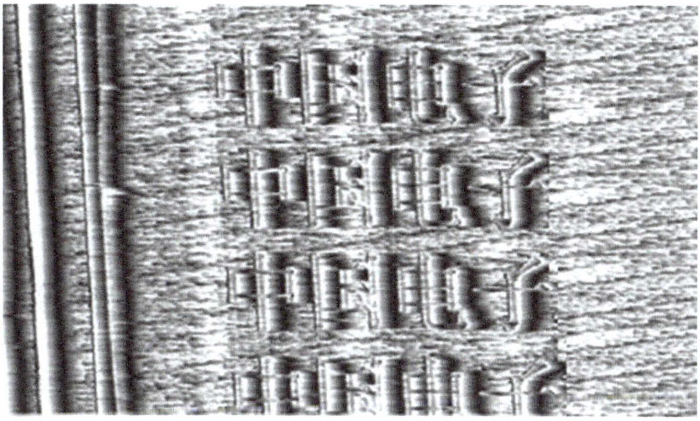

Fig. 6.18 Reproduced image of laptop LCD

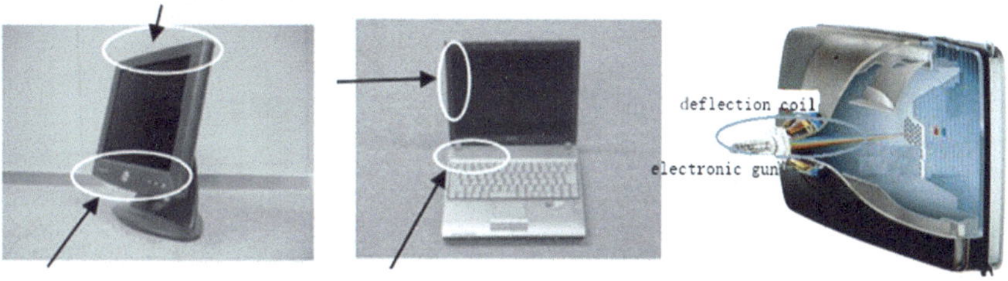

Fig. 6.19 The main radiation part of all kinds of displayers

working circuits and signal lines. Near-field probe can be used to test radiation intensity around these locations and find radiation source.

It is necessary to take appropriate anti-leakage measures for electromagnetic radiation of display in process of designing and producing. Different protection levels have different costs. Therefore, for specific application scenarios, efforts need to be made to strike a balance between security and cost, and select products or protection measures that meet requirements of TEMPEST standard.

For example, security level of NATO TEMPEST standard is divided into three levels: A\B\C (SDIP27 LEVEL A\SDIP27 LEVEL B\SDIP27 LEVEL C). Figure 6.20 lists notebook laptops with different security levels that meet this standard.

6.3.3 *Input Device—Keyboard*

Keyboard is a common input device, including PS/2 keyboard, USB keyboard, notebook portable computer keyboard, and wireless keyboard. As early as 1990, conducted emissions in serial cables were found. PS/2 keyboard transmits key code to

Security Level	Type	Picture
NATO's TEMPEST standard Safety class A SDIP27 LEVEL A	SIL720-L530 notebook	
NATO's TEMPEST standard Safety class B SDIP27 LEVEL B	SIL788-L530 notebook	
NATO's TEMPEST standard Safety class C SDIP27 LEVEL C	SIL784-L530 notebook	

Fig. 6.20 Portable computer products with different security levels

host through serial bus, so PS/2 keyboard will inevitably generate direct electromagnetic radiation.

Bulky documents illustrate that the keyboard is a form of electronic device which is potentially threatening to the information security.

There is a fine example to verify that keyboard is the source of electromagnetic leakage from Safety and Cryptography Laboratory at the Federal Institute of Technology in Lausanne. Researcher Martin Vuagnoux and Sylvain Pasini intercepted electromagnetic radiation of wired keyboard and reproduced keystrokes of keyboard. Four different attack methods were discovered, and various wired keyboards and notebook keyboards produced since 2000 were successfully realized.

As the example of PS/2 keyboard, the feature of electromagnetic information leakage of the keyboard is checked by reconstruction technique of falling edge of transformation.

Keyboard sends scan code to host in serial bus according to protocol. Figure 6.21 shows scan code transmission protocol of PS/2 keyboard, consisting of 11 bits, 1-bit start bit, 8-bit scan code, 1-bit parity bit, and 1-bit stop bit, respectively, and host reads data at falling edge of clock. Figure 6.22 is its interface circuit.

The rising edge time of this circuit is 2 μs and the falling edge time is 200 ns, falling edge is steeper than rising edge with more high-frequency components, and electromagnetic wave will be caused and radiated into space by equivalent antenna. So, electromagnetic leakage energy of the falling edge is greater than that of the rising edge according to Maxwell equation and Fourier transform.

Data time series of bus is shown in Fig. 6.23a when the keyboard E key is stroked, the transmission signal of the clock data line is shown in Fig. 6.23b, the transmission data is 0 × 24, and the total is 11 bits (00010010011B) which is added the start bit and parity bits and stop bit. An electromagnetic leakage signal is generated at falling edge of data and clock, as shown in Fig. 6.23c.

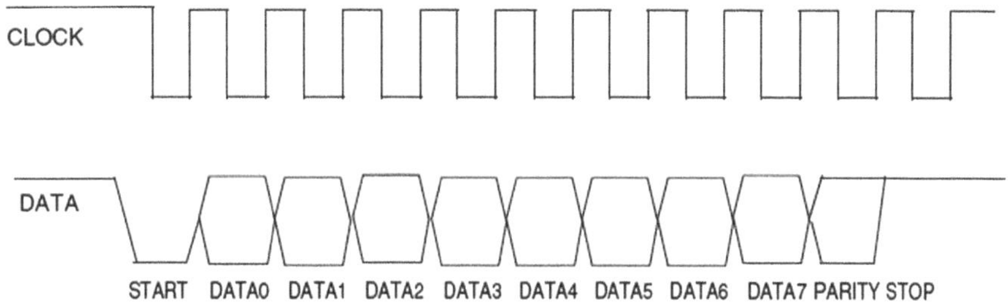

Fig. 6.21 PS/2 protocol format chart

Fig. 6.22 Interface circuit of keyboard

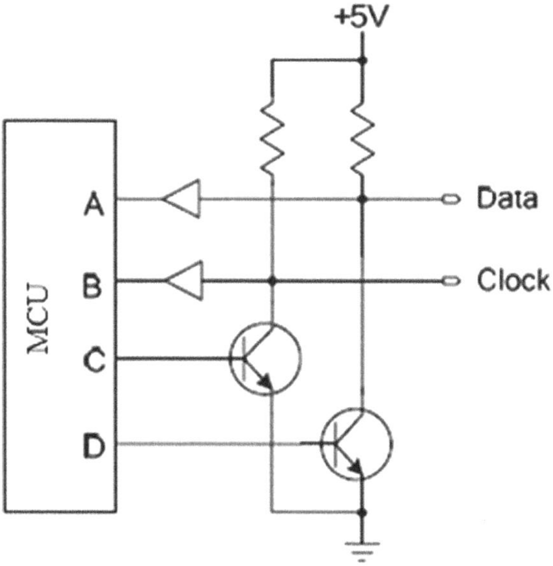

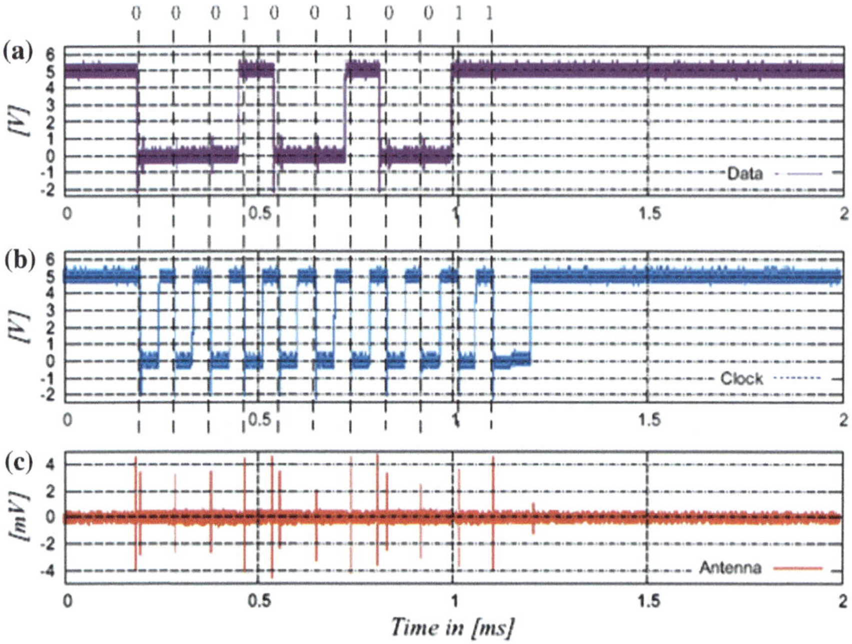

Fig. 6.23 Keyboard electromagnetic signals

The electromagnetic radiation characteristics of the data and clock can be found by analyzing Fig. 6.23c, as shown below:

(1) Both falling edges of clock and data lines produce strong electromagnetic radiation. Therefore, falling edge is the key point of electromagnetic leakage.
(2) Falling edge of clock and falling edge of data do not overlap, so a "double line" occurs, with falling edge of the data leading and falling edge of the clock being followed. The "double line" represents data having a transition from "1" to "0". The data at this moment is "0".
(3) In the case of where the "double line" does not appear, voltage level has an influence on electromagnetic radiation generated by falling edge of clock. When data is "1", high level, the electromagnetic radiation is relatively strong. Therefore, when clock radiation intensity is large, the data is "1", and when clock radiation is small, the data is "0".

Therefore, keyboard scan code can be read from the intercepted electromagnetic signal in Fig. 6.23c as "00010010011B", which is completely consistent with the data transmitted in Fig. 6.23a, b, and the data is completely weighted reconstruction. This example is a good proof that the keyboard is a source of leakage of electromagnetic information. The example proves the keyboard a part of the leakage source of electromagnetic information.

In addition to wired keyboards, wireless keyboards are also widely used. The types of wireless keyboards mainly include infrared communication keyboards and radio communication keyboards. The latter is more commonly used, and its frequency points are 330 MHz, 455 MHz, 900 MHz, and 2.4 GHz.

Compared with wired keyboards, wireless keyboards communicate via electromagnetic waves, which are intentional radiation, but at the same time accompanied with unintentional leakage. Features of electromagnetic information leakage for wireless keyboards are the following:

(1) Electromagnetic leakage of wireless keyboard is generated when it exchanges information with host.
(2) Electromagnetic radiation intensity generated by keyboard decreases as transmission distance increases.
(3) Spectrum line of keyboard is related to the key pressed.
(4) The maximum radiation intensity of each key is different.

The signal frequency spectrum caused by four keys of W, H, Z, and H is shown in the following Fig. 6.24.

It can be found that the spectrum of these images has significant differences. Since each button has a unique scan code, spectrum and maximum radiation intensity of each button are also different.

The information exchange between wireless keyboard and host is a signal transmitted through antenna. Therefore, electromagnetic compatibility of low-radiation

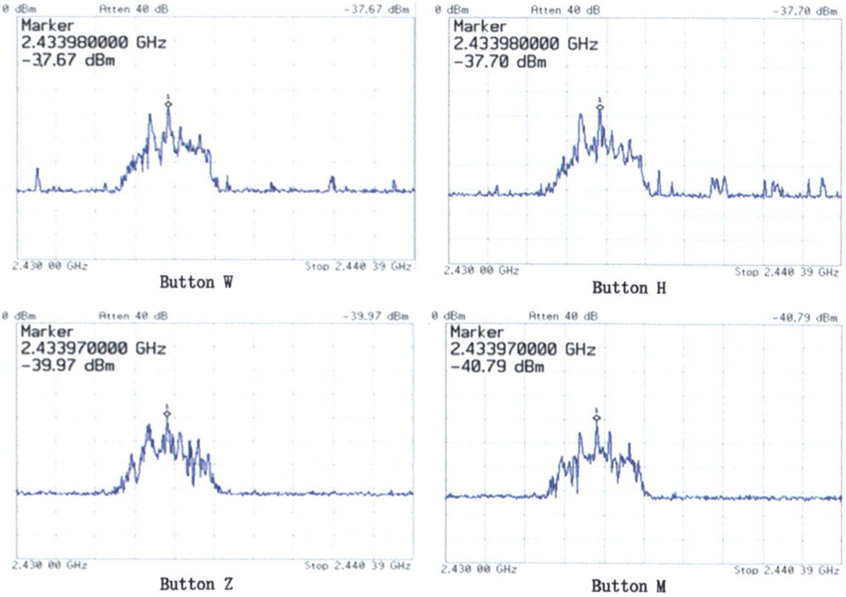

Fig. 6.24 Electromagnetic radiation diagram of wireless keyboard (test distance is 0.5 m, frequency: 2.43–2.44039 GHz)

technology, such as source suppressing, shielding, filtering, etc., cannot be used to protect information security of wireless keyboard. Because this kind of technology is mainly to suppress electromagnetic emission, it is contrary to mechanism of wireless keyboard. In this case, noise interference and information encryption should be used, without affecting normal operation of wireless keyboard.

6.3.4 Coding Technique—Signal Transmission Line

Computer components are linked by the guideline, such as the control line and data line and the address or the order line shown in Fig. 6.3, in general, all of these are called as the signal line. The signal transmission line is the source that is most prone to electromagnetic information leakage. The data line is divided into sequential line and parallel lines, in which the electromagnetic radiation caused by the sequential transmission is more easily interrupted and captured and read compared with parallel transmission. So, electromagnetic information leakage of sequential signal line and its interface is more dangerous.

For instance, LCD display is belonging to low-radiation equipment relatively compared with a CRT display. Because LCD display needs to connect hosts with the signal line which is the main leakage source, LCD display exits the electromagnetic leakage dangers as same as a CRT display, and it is necessary that electromagnetic information leakage protection should be adopted in various kinds of signal line.

Shared connection line of hosts and display is VGA interface line which is the widely applied interface type on display card and configuration in most display card. There are 15 needles in VAG interface and three rows and five holes in each row. Analog signals of the red, the blue, the green, and synchronizing signals (horizontal and vertical) are transferred, analog signals of three primary color transferred are 0–0.7 V, high and low levels of synchronizing signals are 0–0.5 V/2.4–5.5 V, respectively. The digital interface line is adopted in the industry because VGA interface line can't meet the needs of process features for flat-panel display with the development of flat-panel display, especially the emergence of high-end display. Commonly, there is FPD LINK interface of NEC for digital interface line in portable computer. There are three kinds of standards in desk computer including VESA Plug & Display (P&D), VESA Digital Flat Panel (DFP), and Digital Visual Interface (DVI).

Digital coding technology applied in digital interface line is a direct consequence of the electromagnetic informance security, if the coding technology is incorrectly chosen, and the leakage signal is easily reconstructed and reproduced compared with the traditional CRT display connection line. The choice of coding should follow the principle of "anti-leakage, anti-interception, anti-recurrence", which can reduce the signal change or keep signal changes to follow the same pattern during data transmission to reduce the leakage prevention and interception. For example, Transition Minimized Differential Signaling can achieve this goal.

TMDS-coded cable is similar to the FPD-LINK interface. It consists of four twisted pairs, three of which are used to transmit three primary colors, and the fourth group is used to transmit clock synchronization signals. The TMDS encoding converts each set of color 8 bits of value *d7d6…d0* into a 10 bit of *q9q8…q0*. The specific steps are shown in Fig. 6.25.

The conversion step consists of the following two steps:

(1) Transition from *d7d6…d0* to *q8q7…q0* (8 bit → 9 bit).

If the purpose of the step is reduced, it changes number as little as possible and narrow the distance of hamming.

For instance, (*d0*, *q0* on the right).

10101010 ——> 0 11001100,   01010101 ——> 1 00110011,

00000000 ——> 1 00000000,   11111111 ——> 0 11111111.

(2) Transition from *q8q7…q0* to *q9q8—q0* (9 bit → 10 bit).

Q9 is located at limiting the number of differences of "1" and "0" transmitted for the purpose of direction current balance.

For instance, (*d0*, *q0* in left).

00000000, 00000000, 00000000, 00000000, 00000000, . . . ——>

0000000010, 1111111111, 0000000010, 1111111111, 0000000010

1111111111, 0000000010, 1111111111, 0000000010,

0000000010, 1111111111, 0000000010, 1111111111, 0000000010

1111111111, 0000000010, 1111111111, 0000000010,

. . .

11111111, 11111111, 11111111, 11111111, 11111111, . . . ——>

0000000001, 1111111100, 1111111100, 0000000001, 1111111100

0000000001, 1111111100

0000000001, 1111111100, 1111111100, 0000000001, 1111111100

0000000001, 1111111100

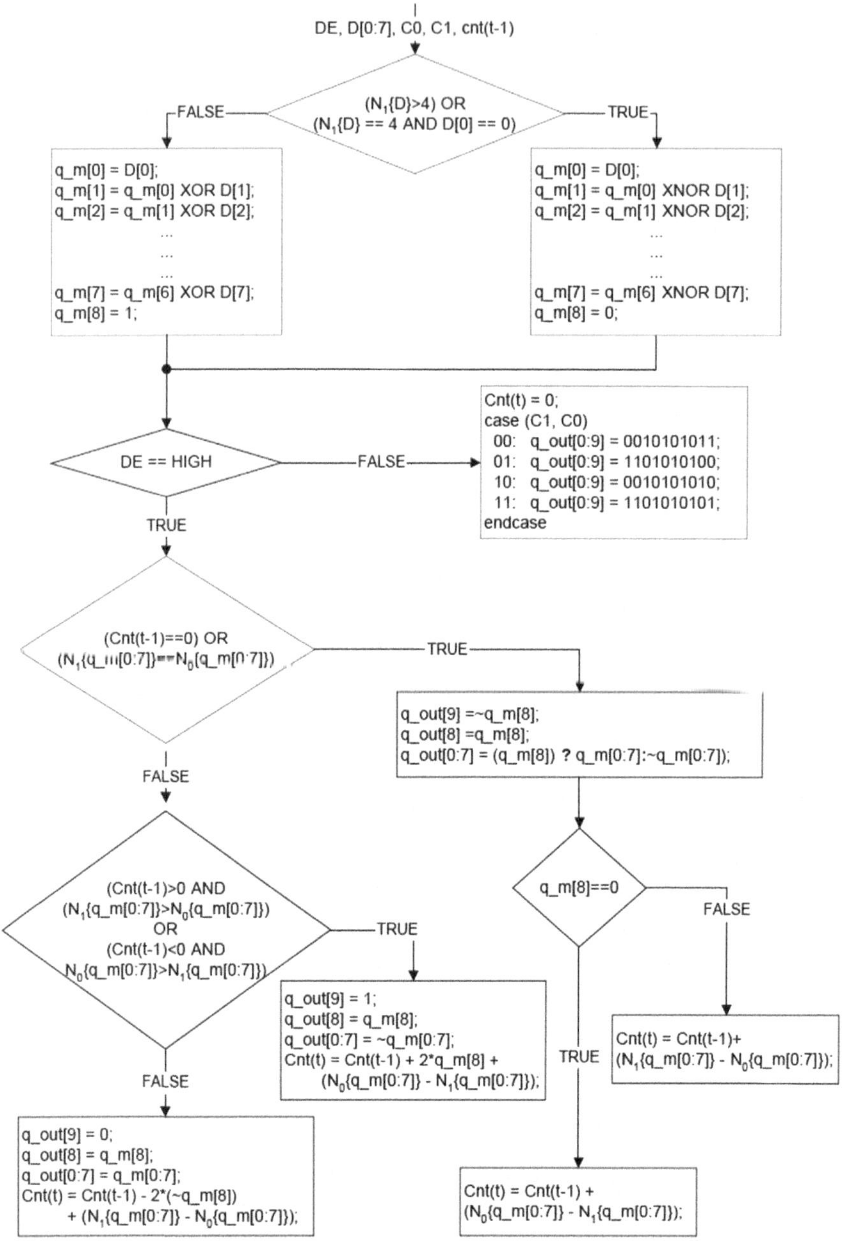

Fig. 6.25 TMDS coding calculation process

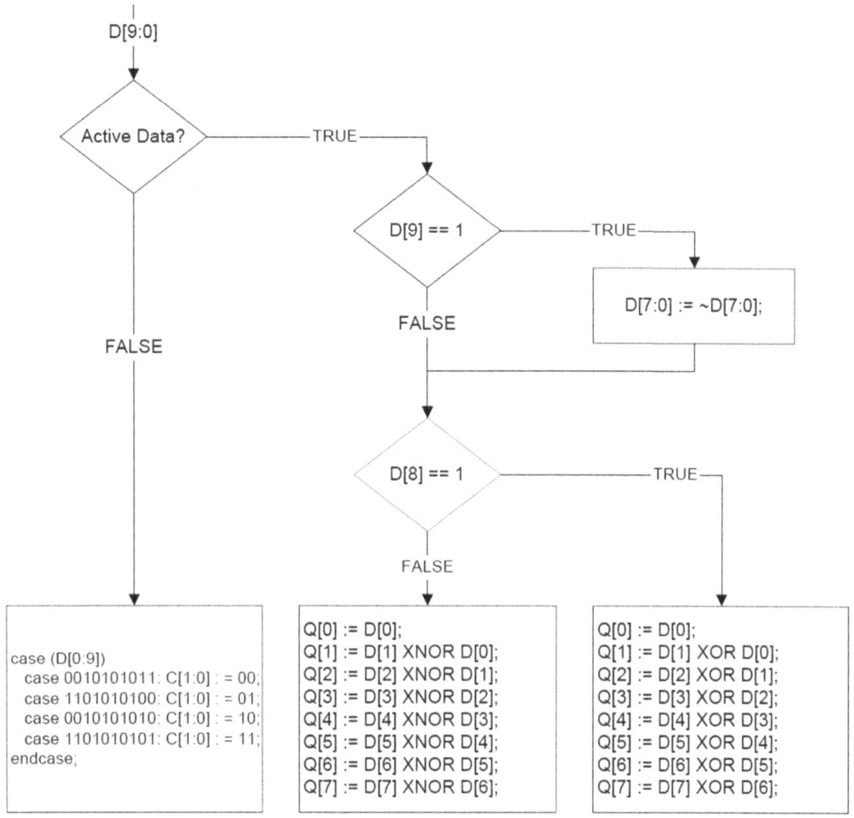

Fig. 6.26 TMDS decoding process

The decoding process of TMDS is shown in Fig. 6.26.

The following is a comparison of a series of intercepted and reproduction instances [60, 61]. The first group adopts conventional coding techniques to reduce radiation changes and the second group adopts TMDS coding techniques to reduce radiation changes.

(1) Conventional coding technology, coding scheme, and display effect are shown in Fig. 6.27, and the effect of its replication is shown in Fig. 6.28.
(2) TMDS coding technology. The encoding scheme and display effect are shown in Fig. 6.29. The effect of the replication scheme is shown in Fig. 6.30.

line	description	foreground		background	
		RGB	signal	RGB	signal
1	black on white	00 00 00	000000x 0x00000 xxx0000	ff ff ff	111111X 1X11111 xxx1111
2	maximum contrast	a8 50 a0	010101x 0x01010 xxx1010	00 00 00	000000x 0x00000 xxx0000
3	maximum contrast (gray)	a8 a8 a8	010101x 1x10101 xxx1010	00 00 00	000000x 0x00000 xxx0000
4	minimum contrast	78 00 00	001111x 0x00000 xxx0000	00 f0 00	000000x 0x11110 xxx0000
5	minimum contrast	78 60 00	001111x 0x01100 xxx0000	30 f0 00	000110x 0x11110 xxx0000
6	minimum contrast (phase shift)	70 70 00	001110x 0x01110 xxx0000	38 e0 00	000111x 0x11100 xxx0000
7	text in most significant bit, rest random	—	r1rrrrx rx1rrrr xxx1rrr	—	r0rrrrx rx0rrrr xxx0rrr
8	text in green two msb, rest random	—	rrrrrrx rx11rrr xxxrrrr	—	rrrrrrx rx00rrr xxxrrrr
9	text in green msb, rest random	—	rrrrrrx rx1rrrr xxxrrrr	—	rrrrrrx rx0rrrr xxxrrrr

Fig. 6.27 Coding scheme and display effects of conventional coding technology

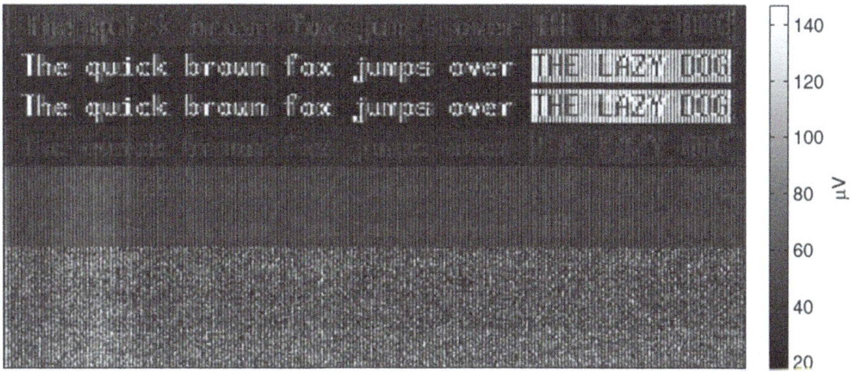

Fig. 6.28 Reconstruction effects of conventional coding

line	description	foreground RGB	background RGB
1	black on white	00 00 00	ff ff ff
2	maximum bit transition contrast	00 00 00	aa aa aa
3	half bit transition contrast	00 00 00	cc cc cc
4	balanced word, max contrast	10 10 10	55 55 55
5	minimum signal contrast	ff 00 00	00 ff 00
6	low nybble random	0r 0r 0r	fr fr fr
7	text in msb, rest random	—	—
8	text in green two msb, rest random	—	—
9	text in green msb, rest random	—	—

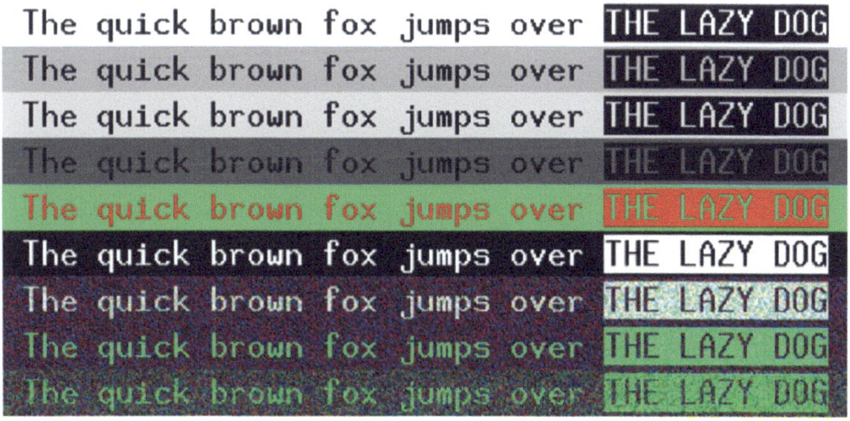

Fig. 6.29 Coding scheme and display effects of TMDS coding technology

Fig. 6.30 Reconstruction effects of TEMDS coding

The following conclusions are drawn by comparing the case studies of the core components of the computer system such as the CPU, display, keyboard, and signal transmission lines:

(1) Reduce the number of logic-level transitions to reduce the risk of leakage of electromagnetic information during signal transmission.
(2) Designing and selecting a reasonable coding technology before data transmission reduces the number of digital changes, and the anti-recovery effect of anti-interception is obvious.

The above conclusions further confirmed the effectiveness of the electromagnetic information leakage prevention strategy proposed in Chap. 5.

Chapter 7
Data Process of Electromagnetic Information

Importance of information security is determined by the value of information. Signal is the carrier of the information and data is its expression.

The information is transformed into the form of signal in transmission and also need to be processed. From the perspective of anti-electromagnetic information leakage, the purpose of signal processing is transformation, improved reliability, and transmission efficiency. The common methods of signal processing are Fourier transformation, wavelet transformation, filter technology, and so on.

Information is stored and processed in the form of data. New information is provided by the processed data. Considering anti-electromagnetic information leakage, data processing increases difficulty of intercepting, reconstructing, and reproducing leaked information for eavesdropper, thereby increasing security and reliability.

The following will introduce information and its transmission, followed by introduction of methods of signal processing and data processing. Finally, give comparison and summarization of the methods from the angle of system-level protection strategy.

7.1 Information

Information is of universality and abstractness. The relations of information, data, and the signal are shown in Fig. 7.1.

Information theory is the theory which is researched on information measure, channel capacity, information distortion rate, and signal channel encoding [10]. The constitution is shown in Fig. 7.2.

(1) Information Measurement

The primary problem of information theory is to measure information. The measurement is to treat information source as a stochastic process with a certain prior probability. The entropy is regarded as a measurement of information source, represented by H, which is defined as probability of certain information occurrence,

T. Liu and Y. Li, *Electromagnetic Information Leakage and Countermeasure Technique*,
https://doi.org/10.1007/978-981-10-4352-9_7

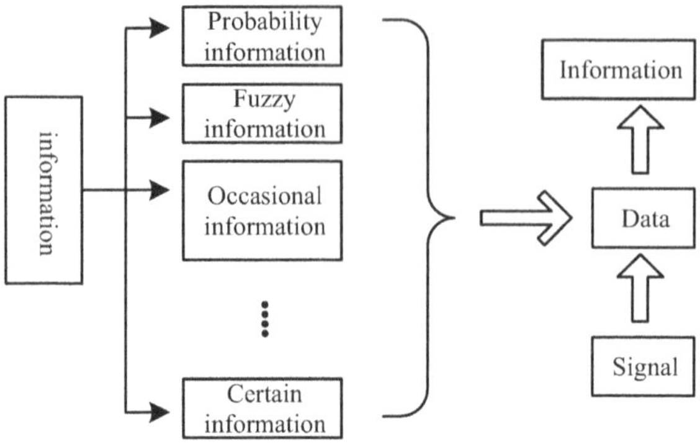

Fig. 7.1 Relation of information, data, and signal

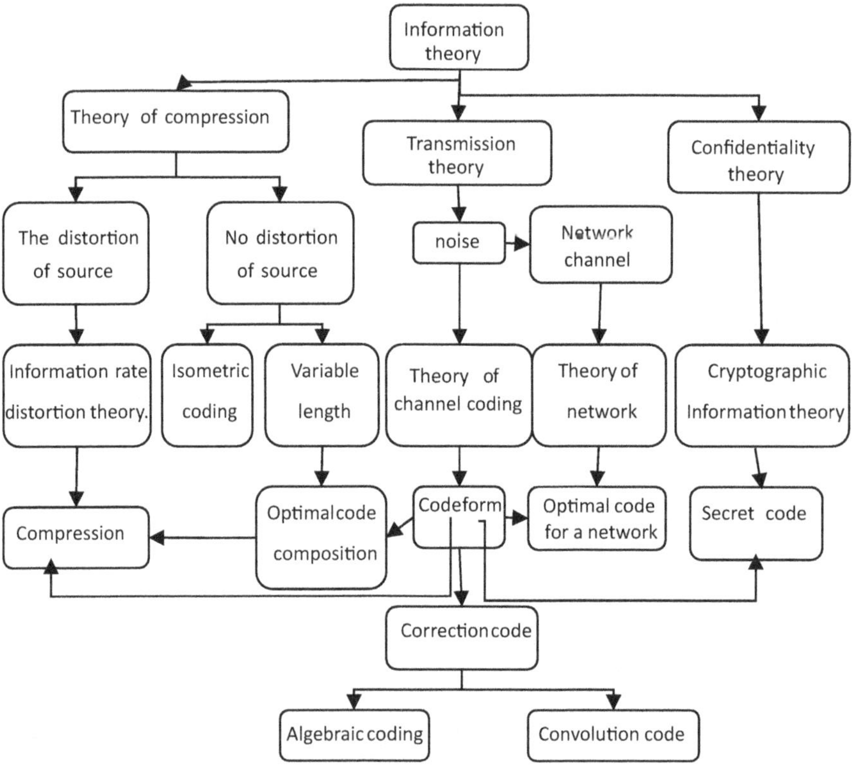

Fig. 7.2 The composition of information theory

including probability information, fuzzy information, sporadic information, determined information, etc.

$$H(x) = -\sum_{i=1}^{n} p(x_i) \log_b p(x_i) \tag{7.1}$$

In this formula, H(x) is entropy; p is probability function of certain information x; b is the bottom of logarithm, when b = 2, the unit of entropy is in it.

(2) Coding Theorem of Undistorted Information Source

If the information source code rate is no less than the entropy of the information source, there are undistorted information source and vice versa. The brief introduction is as the following:

If $R_1 \geq H$, there is undistorted information source, in which R_1 is information source code rate and H is entropy of the information source.

Coding theorem of undistorted information source is the theoretical basis of information source compressed encoding which is called Shannon's first principle also.

For example, English letters and space characters have a total of 27 symbols, which are transmitted in code, and each symbol requires 5 bits. However, it is found that the actual English letter source entropy is about 1.4 bits/symbol. According to Shannon's first theorem, there is a certain source coding method, and each letter can be transmitted in 1.4 binary character symbol without distortion. This can significantly improve transmission efficiency.

(3) Noisy Channel Coding Theorem

If the Rate of Information Throughput (RIT) is less than channel capacity, an encoded mode can be found in which the transmission error is arbitrarily small when coded sequence is long enough; otherwise, there isn't encoding with arbitrary minor error. The brief introduction is as the following:

If $R_2 \leq C$, there is encoding with arbitrary minor error, in which R_2 is information transmission rate and also called channel coding rate. C is channel capacity.

Noisy channel coding theorem is the theoretical basis of channel coding (Shannon's second theorem).

(4) Theory of Signal Rate Distortion

In many cases, the information allows transmission with a certain amount of error and does not need accurate transmission. Thus, transmission efficiency can be improved and communication cost can be decreased under the premise to obtain enough information. In order to achieve effectiveness of distortion-limited transmission, the goal is to minimize code rate of encoder under the condition that distortion rate is lower than a certain value.

This is a theoretical problem of lossy compression coding optimization based on theory of information distortion rate. The core of it is distortion-limited source coding theorem. The brief introduction is as the following:

For any distortion measure D, encoding B can be found so long as the code word is long enough, when the code rate of information source encoding R1 > R(D), the average distortion of decoding is less than D. Otherwise, if the code rate of information source encoding R1 < R(D), there isn't the encoding in which the average distortion is less than D.

R1 is information source code rate, R(D) is information rate distortion function, and B is the minimum code bit number needed for each information source symbol in order to satisfy the distortion rule (the average distortion is less than D).

Distortion source coding theory is Shannon's third theorem which is the theory basis of loss compression coding.

For instance, the discrete information source is 16 characters with equal probability. If undistorted coding is required, each character needs four binary symbols, that is, the average code length is 4 bits. If the distortion is allowed, the average code length can be decreased. If the average distortion is required to be less than 1/2, a maximum of 1/2 error is allowed when code sequence is decoded. Obviously, the first seven symbols of the information source can be encoded, respectively, the last nine symbols of the information source can be programmed as the same code word (the eighth) to transfer, when it is decoded on the receiving terminal, the first seven code word received is corresponded one to one with the first seven symbols of the receiver symbol table without decoding error. When the eighth code word is retrieval, anyone can be random decoded code word from the 8th to 16th in the retrieval terminal table. The accuracy rate of total decoding is $7/16 + 1/16 = 1/2$. So, the error rate is 1/2 also, the encoding distortion rate is 1/2. If the optimal binary encoding can be applied to calculate (for instance, Huffman encoding), the average code length of coding is up to 2.25 bits which are less than 4 bits. But, in theory, the minimum code length of information source encoding is decided by R(D) in order to realize the 1/2 of error rate in the same information source, which is 1.05 bits.

(5) Information Security Code

Information security code is encrypted transmission information and then transformed through the channel coding for the purpose of improvement of the information security. Information system encryption theory is the basis of information security password.

A common security transmission method is to encrypt plaintext, using an encryption algorithm such as DES, to obtain its corresponding ciphertext and then transmit it through channel.

According to information transmission theory, information transmission model is composed of information source, channel, and destination.

Information source is divided into discrete source and continuous source, as shown in Fig. 7.3.

As same as the classification of the information source, in general, the information channel is divided into discrete channel and continuous channel according to the relationship between statistical property and time, statistical property of input and output signal, number of customers, the influence from output end to input end, and so on, respectively, as shown in Fig. 7.4.

<table>
<tr><td rowspan="6">Source cf information</td><td rowspan="4">Discrete information source</td><td rowspan="2">Discrete memoryless information source</td><td>Single signal in discrete memoryless information source</td><td>Independent event</td></tr>
<tr><td>The symbol sequence is discrete without memory information source</td><td>A series of sequences of symbols, English: a sample space of 26 letters, Spaces, and punctuation marks.</td></tr>
<tr><td rowspan="2">Discrete memory information source</td><td>A symbol sequence has a memory information source</td><td>Interdependence between symbols, English: words, grammar and rhetoric</td></tr>
<tr><td>The symbol sequence of markov information source</td><td>There is a strong correlation between the symbols and the adjacent symbols, and the correlation with the former is weakened。</td></tr>
<tr><td rowspan="2">Source of continuous source</td><td>Continuous time and amplitude</td><td>Speech signal, thermal noise signal, TV image signal</td><td rowspan="2">Through sampling and quantification, the continuous source is transformed into discrete source to be processed。</td></tr>
<tr><td>Continuous time or amplitude</td><td></td></tr>
</table>

Fig. 7.3 The classification of the information source

Besides, information is divided into noiseless channel and noisy channel. Usually, channel noise is equivalent to noise and interference from other parts of the system. There are two kinds of background noise in a communication system which are additive noise and multiplicative noise. In general, background noise is additive and the fading is multiplicative. Ideal Additive Gaussian White Noise channel (AWGN) is mostly researched in information theory. The cause of research for Gaussian noise is its universality and ease of handling.

7.2 Signal Processing Methods

The number, frequency, power, polarization, working time, mechanism, and so on of electromagnetic radiation source determine the frequency band, space, and time distribution of electromagnetic information.

The processing of electromagnetic information usually analyzes frequency, time, and spatial characteristics of the intercepted signal. Conventional detection methods for electromagnetic information processing methods include Fourier transform,

channel	The relationship between statistical property with time	channel of constant reference (Stationary channel)	Statistical properties do not change with time	
		Variable reference channel	Statistical properties change over time	
	statistical property of input and output signal	Discrete channel (digital channel)	Input and output signal sets The values are discrete	Average mutual information
		Continuous channel (Analog channel)	Input and output signal sets The values are continuous	The capacity of the shannon channel
	Number of customers	Single -user channel (Signal at both ends)	There is only one input and output end	
		Multi -user channel (Signal at multiple ends)	There are more than two users at least one end of the input and output. such as Satellite communication.	
	the influence from output end to input end	Non feedback channel		
		Feedback channel		

Fig. 7.4 Classification of channels

wavelet transform and filtering techniques [63–66]. In addition, there are methods that are unique to the detection mechanism, such as stochastic resonance and chaotic methods [75].

7.2.1 Fourier Transform

Fourier transform can transform the space-time signal into a frequency signal. Functions (sequence or vector) generated from Fourier transform include the spectrum signal which can clearly reveal frequency construction of the information. So, it is also called as spectrum analysis, and widely applied in many fields such as signal processing, physics, probability theory, statistics, cryptography, acoustics, optics, and so on.

1. Fourier Series

The Fourier series is the original form of Fourier series development. Any continuous time period function can be decomposed into the superposition of many different frequency sinusoidal signals by Fourier series. Meantime, it reveals the essence of analog signal which is sum of sinusoids of different frequency, amplitude, and phase positions. Actually, Fourier series is one kind of Fourier transform of continuous time and discrete frequency.

For continuous period function X(t) satisfying Dirichlet conditions, its expansion of Fourier series is

$$x(t) = \sum_{k=-\infty}^{\infty} X(jk\Omega_0)e^{jk\Omega_0 t} \tag{7.2}$$

In the expression, Ω_0 is interval of frequency spectrum line of signal, which is angular frequency.

From the expression (7.2), a periodic function is likely to be the linear combination of infinite multiple harmonic functions and each harmonic function is corresponded to a frequency component in which the frequency of one harmonic component is the reciprocal value of the function periods and the frequency of all other high harmonic components is the integer multiple of base frequency.

Fourier series is of significant meaning to electromagnetic information analysis. When processing the information, the information facilities always work at a certain working frequency and it is closely related between processing instruction cycle and working frequency in the facilities internally. In other words, it is an internal relationship between the working frequency of the information facilities and its processed information. The working frequency is sequence pulse and its signal form is a periodic square wave. According to Fourier series, the square wave can be expanded into the superposition of the odd harmonic component, that is,

$$f(t) = \frac{4A}{T}(\sin \omega_0 t + \frac{1}{3}\sin 3\omega_0 t + \cdots + \frac{1}{n}\sin n\omega_0 t) \tag{7.3}$$

In the expression, A is the amplitude, T is period, and ω_0 is fundamental angle frequency; n = 1, 3, 5, ….

It means that the working frequency of the facilities can be obtained once the harmonic component of an information device timing pulse is intercepted and captured and then it is possible that the processing information can be analyzed by means of data reconstruction and information retrieval.

2. Discrete Fourier Transform (DFT)

Usually, the data are discrete and finite because the digital signal is processed by the information facilities when the generated electromagnetic information is analyzed. The forms of expression of Fourier transform are different according to various time-frequency characters in the processed signal. Discrete Fourier Transform (DFT) is applied during the processing of discrete data signal. The digital signal processed by the information facilities can be expressed by a finite discrete sequence. In fact, the sequence is a function with discrete time and discrete frequency, as shown in Fig. 7.5. $x(n)$ is time function of the signal, $x(k)$ is frequency function of signal, sampling period is T, and time function period is $Tp = NT$.

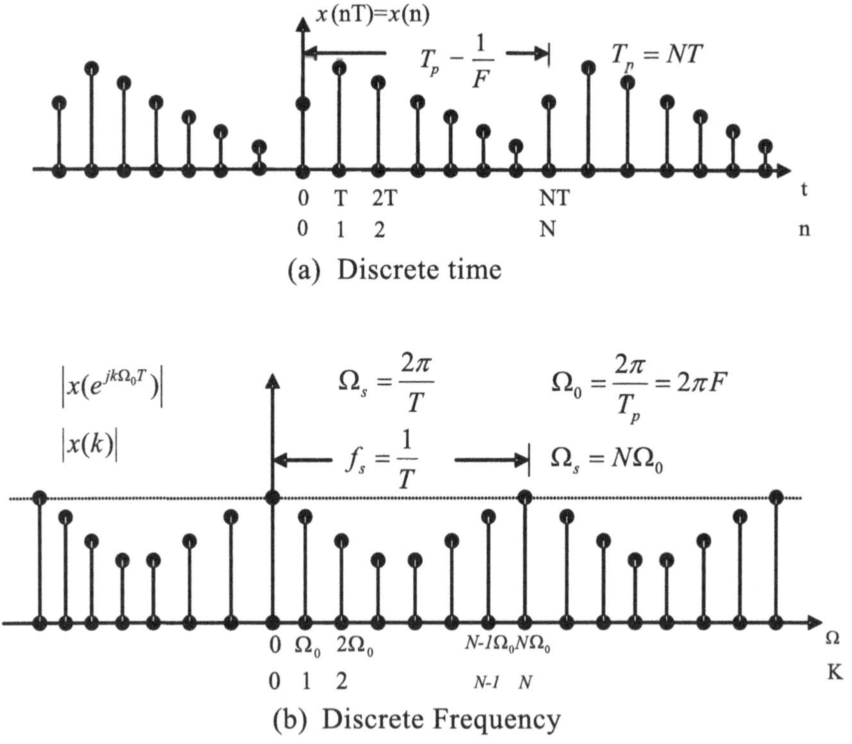

Fig. 7.5 Discrete time and discrete frequency signals

Time function $x(n)$ and frequency-domain function $X(k)$ of the discrete signals can be shifted to each other. For a finite length $x(n)$ of length N, there is no zero value in N-point only $n = 0$ *to* $n = N - 1$. The rest is zero. The formula of discrete Fourier transform is shown below:

$$X(k) = \sum_{n=0}^{N-1} x(n)e^{-j\frac{2\pi}{N}nk} \tag{7.4}$$

$$x(n) = \frac{1}{N}\sum_{k=0}^{N-1} X(k)e^{j\frac{2\pi}{N}nk} \tag{7.5}$$

Formula (7.4) is a direct transformation of Fourier that the signal is transformed from the time-domain function $x(n)$ to frequency-domain function $X(k)$.

Formula (7.5) is the inverse transformation of Fourier that the signal is transformed from the frequency-domain function $X(k)$ to time-domain function $x(n)$.

$W_N = e^{-j\frac{2\pi}{N}}$ is defined by the transformation factor of Fourier.

From the above analysis, we know that the frequency domain must be periodic if its time domain of the signal is discrete. Conversely, the frequency domain must be discrete if its time domain of the signal is periodic, as shown in Table 7.1.

In signal processing, the Discrete Fourier Transform (DFT) plays a decisive role. Signal correlation analysis, filtering, spectrum estimation, etc. can all be realized by discrete Fourier transform.

3. Fast Fourier Transform (FFT)

Discrete Fourier Transform (DFT) provides an analysis method of digital signal but the calculation is very large. When performing a discrete Fourier transform on a sequence of length N, N^2 complex multiplications and N (N − 1) complex additions are required. As the sequence length N increases, its computational complexity will increase greatly.

Fast Fourier Transform (FFT) subtly utilizes the periodicity and symmetry of W_N factor to decompose the long sequence, to be computed, into a number of short sequences, and realizes fast calculation of discrete Fourier transforms by multistage butterfly calculation.

For a finite length sequence X(N) of length N, the N-point FFT operation requires only (N/2) $\log_2$N complex multiplication operations, and the discrete Fourier transform operation requires running N^2 complex multiplication operations.

For instance, when N = 1024,

Table 7.1 Time-domain characters of discrete time and discrete frequency signals

Time-domain signal	Frequency-domain signal
Discrete	Cyclical
Cyclical	Discrete

$$N = 1024, (1024/2) \times \log_2 1024 = 512 \times 10 = 5120$$

The fast Fourier transform requires 5120 complex multiplications. The discrete Fourier transform requires $1024^2 = 1,048,576$ times 5120/1,048,576 = 4.88%. The operation speed is increased by 20 times.

The operation speed is largely improved by Fast Fourier Transform (FFT), so it is widely applied in digital signal processing. The algorithm is used in not only various kinds of signal processing software, but also solidified hardware by circuit design and the processing chip integration.

4. Short-Time Discrete Fourier Transform (STFT)

To overcome the faultiness of Fourier transform without local analysis capacity, short-time Fourier transform divides the signal into a series of small time intervals, and each time interval is analyzed by Fourier transform to determine frequency of the time interval. The short-time Fourier transform is expressed as

$$S(\tau, \omega) = \int_{-\infty}^{\infty} f(t)g(t - \tau)e^{-j\omega t}dt \tag{7.6}$$

where f(t) is the signal to be transformed, g(t) is the window function, commonly known as Hann window or Gaussian window centered around zero, S(τ, ω) is essentially the Fourier transform of f(t)g(t − τ), a complex function representing the phase and magnitude of the signal over time and frequency.

STFT also has a disadvantage that when window function g(t) is determined, the shape of the rectangular window is determined. τ and ω can only change the location of the window in the phase plane but cannot change the shape of the window. Fourier transform is actually an analysis with single resolution ratio. To change the resolution, window function g(t) must be selected again.

7.2.2 Wavelet Transform (WT)

Wavelet Transform (WT) is proposed after Fourier transform, and it is often compared with the Fourier transform. In fact, the Fourier transform can be viewed as a special case of the continuous wavelet transform with the choice of the mother wavelet $\psi(t) = e^{-i\omega t}$. The main difference in general is that wavelets are localized in both time and frequency, whereas the standard Fourier transform is only localized in frequency. Short-time Fourier Transform (STFT) is similar to the wavelet transform, in that it is also time and frequency localized, but there are issues with the frequency/time resolution trade-off. In general, the wavelet transform's multiresolutional properties enable large temporal support for lower frequencies while maintaining short temporal widths for higher frequencies by the scaling properties of the wavelet transform. This property extends conventional time-frequency analysis into time-scale analysis. Wavelet transform is often used for denoising, filtering, edge detection, and so on.

1. Continuous Wavelet Transform (CWT)

The definition of Continuous Wavelet Transform (CWT) is

$$W_f(a, b) = \frac{1}{\sqrt{|a|}} \int_{-\infty}^{\infty} f(x)\overline{\psi\left(\frac{x-b}{a}\right)}dx \tag{7.7}$$

In formula, a is the scaling factor, corresponding with frequency information; b is the shift factor, corresponding with time and space information; $\Psi(x)$ is the wavelet function and also called as basic wavelet or original wavelet; and $\overline{\psi(t)}$ is a complex conjugate of $\psi(t)$.

The process of Continuous Wavelet Transform (CWT) is shown in Fig. 7.6.

Trigonometric function of $sin(x)$ and $cos(x)$ and e^{-jx} can be scaled to form the basement of function space $\{sin(nx), cos(nx)\}$ and $\{e^{-jwx}\}$, so can the mother wavelet scaled and translated to form the basement of function space.

$$\left\{\psi_{j,k}(x) = 2^{-\frac{j}{2}}\psi(2^{-j}x - k)\right\} \text{ and } \left\{\frac{1}{\sqrt{|a|}}\psi\left(\frac{x-b}{a}\right)\right\}$$

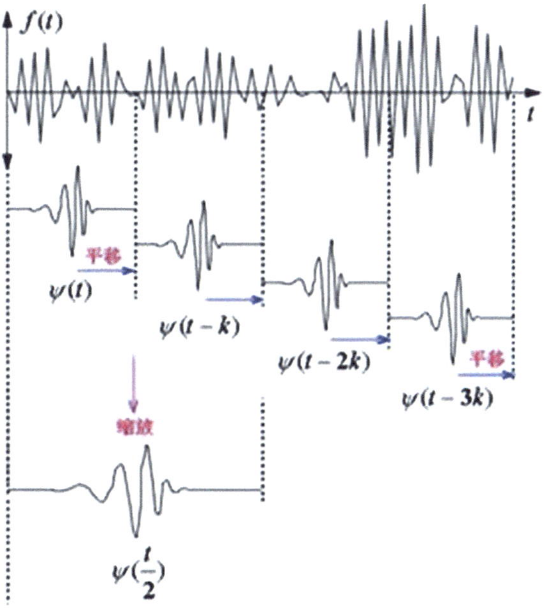

Fig. 7.6 The process of Continuous Wavelet Transform (CWT)

There is difference of the transform kernel between wavelet transform and Fourier transform: the transform kernel of Fourier transform is fixed by virtual exponential function e^{-jwx} (complex trigonometric function); however, the transform kernel of wavelet transform is arbitrary. Because there is an infinite variety of mother wavelet as long as wavelet function $\Psi(x)$ meets the following condition:

(1) Absolutely integrable and square integrable, that is, $\psi \in L^1 \cap L^2$
(2) The positive and negative parts are balanced, that is, $\int_{-\infty}^{\infty} \psi(x)dx = 0, \hat{\psi}(0) = 0$
(3) Meet the allowable conditions, that is, $\int_{-\infty}^{\infty} \frac{|\hat{\psi}(\omega)|^2}{\omega} d\omega < \infty$.

$\hat{\psi}(\omega)$ is Fourier transform of $\Psi(x)$.

Common wavelet function includes Haar wavelet, Mexican Hat wavelet, and Morlet wavelet.

2. Discrete Wavelet Transform

If scaling factor a of continuous wavelet transform is discretized, a binary wavelet transform is obtained; if translation factor b is also discretized at the same time, then a discrete wavelet transform is obtained. In order to accommodate digital signal processing, wavelet transform needs to be discretized.

At first, the scaling factor is discrete, if wavelet function Ψ meets

$$\sum_{k \in Z} |\hat{\psi}(2^k \omega)|^2 = 1 \tag{7.8}$$

Then, Ψ is called as basic binary wavelet. In continuous wavelet transform, if Ψ is basic binary wavelet, let $a = 2^k$, then binary wavelet transform is obtained.

$$W_{2^k} f(b) = \frac{1}{\sqrt{2^k}} \int_{-\infty}^{\infty} f(x) \overline{\psi\left(\frac{x-b}{2^k}\right)} dx \tag{7.9}$$

Then, the shift factor in binary wavelet transform is discrete, let $b = n^{2k}$, then discrete wavelet transform is obtained.

$$W_{2^k} f(n) = 2^{-k/2} \int_{-\infty}^{\infty} f(x) \overline{\psi(2^{-k}x - n)} dx \tag{7.10}$$

Filtering is an effective method to realize the discrete wavelet transform. In fact, this is a signal decomposition method, which is called dual-channel sub-band coding in digital signal processing. The concept of discrete wavelet transform is carried out by filter shown in Fig. 7.7.

In Fig. 7.7, S represents the original input signal. Two signals of A and D are generated by both of the complementary filter. A represents the approximate value

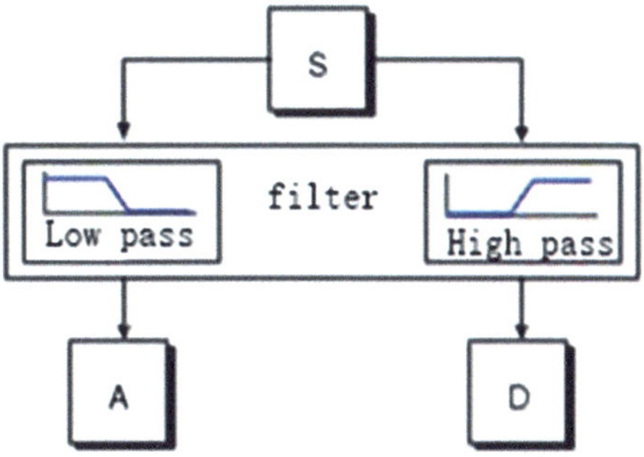

Fig. 7.7 Double channel filter processing

of signal and D represents detailed value of the signal. In many applications, low-frequency portion of signal is the most important, while high-frequency portion is secondary. The approximate value is a coefficient caused by larger scaling factor which represents the low-frequency component of the signal. Meanwhile, the detailed value is a coefficient caused by smaller scaling factor which represents the high-frequency component.

So, the discrete wavelet transform can be regarded as a tree constituted by a low-pass filter and a high-pass filter. The original signal can be decomposed by a pair of filters by which the iteration signal decomposition process can be multilevel decomposed also. Many low-frequency components with lower resolution ratio will be obtained and form the wavelet decomposition tree if the higher frequency component of the signal no longer can be decomposed and the lower frequency component is continuously decomposed. The number of decomposition stages depends on the data to be analyzed and the needs of the user.

Discrete wavelet transform can be applied in signal decomposition. The process of restoring decomposed coefficients to the original signal is called wavelet reconstruction, that is, Inverse Discrete Wavelet Transform (IDWT). Filter-based wavelet transform consists of two processes, filtering and downsampling. And, IDWT consists of two processes, upsampling and filtering. Upsampling is to insert a "0" between two sample data in order to lengthen the signal.

Discrete wavelet transform is widely used in communication, network, information, automatic control, medical engineering, and other fields.

7.2.3 Filter Technology

Filter technology is a technology which can accurately estimate the desired signal from the signal mixed with interference. Filtering theory is based on a certain criterion, using a statistical optimal method to estimate the state of the system from observable signals. That is to say, the purpose of filtering is to estimate implicit state of the system online based on noisy observation.

1. Category of Filter

The filter can be classified into low pass, high pass, band pass, band stop, and so on, according to different filtering frequency bands. In order to pass the desired frequency components and filter out the unwanted frequency components, the corresponding filters can be designed according to the cutoff frequency.

(1) Low-pass filter

Low-pass filter allows low-frequency signal to pass but weakens signal with frequency which exceeds cutoff frequency. For different filters, the signal weakening degree of each frequency is different. There are many various forms of low-pass filter, such as electronic circuit filter, digital algorithm for smoothing data, image blurring, and so on.

(2) High-pass filter

High-pass filter allows the high-frequency signal to pass but weaken (or decrease) signal whose frequency is lower than the cutoff frequency to pass. For different filters, the signal weakening degree of each frequency is different. Sometimes, it is called as low-frequency shear filter. The characters are just the opposite between high-pass filter and low-pass filter.

(3) Band-pass and band-resistance filters

Band-pass filter is a kind of filter which can pass a frequency component during a certain frequency range and attenuate a frequency component to a minimum level in other frequency ranges. Band-pass filter also can be generated by combining low-pass filter with a high-pass filter.

The concept of band-resistance filter is the opposite to band-pass filter. The negation of band-pass filter is equivalent to a band-resistance filter.

2. Signal Process Methods of Filter

According to the signal process methods of filter, there is Kalman filtering, Wiener filtering, Gaussian filtering, and so on.

(1) Kalman Filtering

The problem solved by Kalman filtering is seeking the estimated value in minimum mean square error, of which the feature is calculated by recursive methods. The basic principle of Kalman filtering is that using state model containing signal and noise, the estimation of state variable is updated by the estimated value of previous time and the observation value of current time, thereby obtaining the estimated value of current time.

Kalman filtering is a kind of recursive estimation to work out the estimation value of current status as long as the estimation value of previous status and the observed value of current status are obtained. So, there is no need to record the observed or estimated history information. Compared with most filters, Kalman filter is a kind of pure time-domain filter which does not require the conversion of time frequency like a frequency filter.

Kalman filtering is established on the models of linear algebra and Hidden Markov. The dynamic system can be expressed by a Markov chain which is established on line operator disturbed by Gaussian noise (normal distribution noise). Vector quantity can be expressed by system station with the element of the real number. With the increase of discrete time, this linear operator will act on current state, producing a new state, but will bring in some noise. Meanwhile, some control information of known controller in the system is also added and another linear operator c disturbed by noise can generate the visible output of implicit state.

The model is established under the frame of Kalman filtering in order to estimate the internal state during the observed process by Kalman filtering from a series of observation data with noise, which means definite matrix F_k, H_k, Q_k, or R_k for each step K, B_k also needs to be defined sometimes, as shown in Fig. 7.8.

The circle in Fig. 7.8 is the expressed vector, square is the expressed matrix, star is the expressed Gaussian noise, and covariance matrix is shown at bottom right.

Kalman filtering model assumes the actual state of K moment which is evolved from (K − 1) and meet the following:

$$x_k = F_k x_{k-1} + B_k u_k + w_k \tag{7.11}$$

In the formula, F_k is the state transition model affecting X_{k-1}, B_k is input control model affecting controller vector U_k, w_k is process noise and it is assumed to meet the multivariate normal distribution with zero average value, and Q_k covariance matrix.

$$w_k : N(0, Q_k) \tag{7.12}$$

Moment k, the measurement Z_k of actual state X_k meets

$$z_k = H_k x_k + v_k \tag{7.13}$$

In the formula, H_k is the observation model of which the observation space is mapped by the actual state. V_k is observed noise with zero average value and its covariance matrix is R_k which obeys normal distribution.

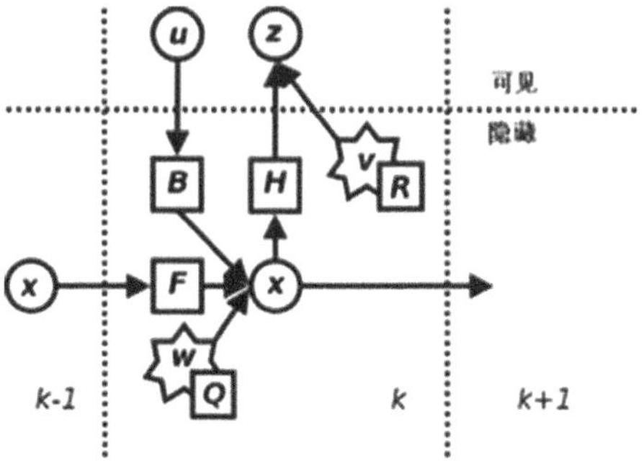

Fig. 7.8 Kalman filter model

$$v_k : N(0, R_k) \tag{7.14}$$

Original state and noise of each moment are independent of each other.

In fact, many real dynamic systems do not exactly match this model. But, Kalman filtering is designed to operate in noisy condition, an approximate match can make this filter work very efficiently.

The operation of Kalman filtering is divided into two stages: prediction and update. At the stage of prediction, filter estimates the current state by the estimation of the previous state. At the update stage, filter utilizes current observation to optimize the prediction value produced in prediction stage, and obtains a new estimate value with more accuracy.

Kalman filtering is a kind of linear filtering. To solve the nonlinear problem, other more complex Kalman filters are proposed, among which the Extended Kalman Filter (EKF) based on local linearization is the most widely used.

EKF is a kind of approximate nonlinear filter of which the principle is to transform the nonlinear system into a linear system and then perform Kalman filtering. So, Extended Kalman Filter (EKF) is suboptimal filtering. Since it was applied in the guidance system of Apollo Project first time, Extended Kalman Filter (EKF) has been already successfully applied in numerous actual systems to solve the nonlinear filtering issue. It assumes that the random part of system obeys Gaussian distribution and the nonlinear system can be well approximated by the linear model based on current state. However, robustness of Extended Kalman Filter (EKF) isn't powerful and easily divergent.

In order to overcome the weakness of extended Kalman filter, a kind of improvement Kalman filter is emerged.

(2) Wiener Filtering

The purpose of Wiener filtering is filtering the interference noise. Normally, the design of filter needs to set a special frequency response, but some of noise will still pass. Wiener filtering realizes the filtering from different perspectives.

Wiener filtering can be classified as deconvolution algorithm (or inverse filtering). The feature of Wiener filtering is to find out optimal filter by the scalar method under an assumption, which in signals and noise have statistically independent expectations. Wiener filtering evaluates the performance of system by minimum mean square error.

The block diagram of the two-dimensional Wiener filter is shown in Fig. 7.9.

Image f(x, y) is linearly blurred by h(x, y), after adding noise n(x, y), and it becomes a degraded graph g(x, y). And, the degraded graph g(x, y) is subjected to Wiener filtering w(x, y) to obtain a restored image $\hat{f}(x, y)$.

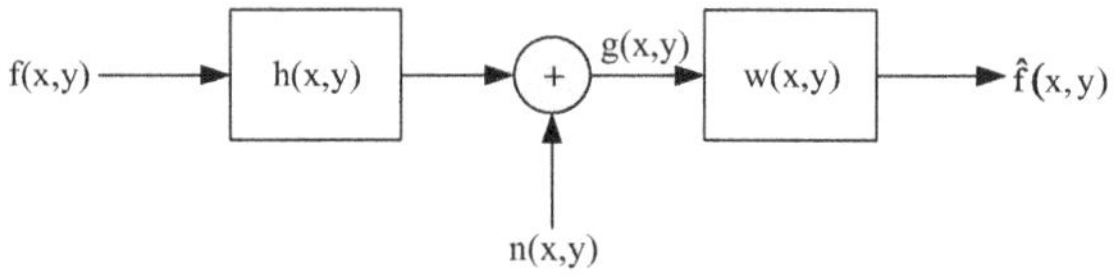

Fig. 7.9 The theory of two-dimensional Wiener filtering

Wiener filtering is widely applied in the field of image restoration due to the good recovery results, low computational complexity, and excellent anti-noise performance. Wiener filter has been continuously improved, and many efficient restoration algorithms have been proposed based on the prototype algorithm.

(3) Gaussian Filter

Virtually, Gaussian filter is a kind of signal filter applied in smoothing treatment of signal which is suitable for elimination of Gaussian noise and widely used in noise reduction process of image treatment. The key to Gaussian filtering is to perform a weighted average calculation of all pixels in the image. The value of each pixel is obtained by weighting average of itself and other pixels in the neighborhood. Gaussian filter adopts the weighted average gray value of neighborhood pixel determined by template to replace the value of each pixel located at the center of the template in a template (or named convolution and mask) scanning image.

Gaussian filter selects weights based on the shape of Gaussian function. Gaussian smoothing filter is very effective for suppression noise that obeys normal distribution. Gaussian function of one-dimensional zero mean is given below:

$$g(x) = e^{-x^2/2\sigma^2} \tag{7.15}$$

In the formula, σ decides the dimension of Gaussian Filter.

Normally, two-dimensional zero-mean discrete Gaussian function is commonly used as a smoothing filter for image processing. The function expression is

$$g(i, j) = e^{-(i^2-j^2)/2\sigma^2} \tag{7.16}$$

Noise is the biggest problem in the digital image processing. To cope with the cumulative transmission of errors, Gaussian filter is applied to obtain the image with higher Signal-to-Noise Ratio (SNR) and reflect the real signal. Thus, Gaussian filter is widely used in image processing due to its significant performances.

First, Gaussian filter is a monotonic function that replaces the initial value of a pixel with a weighted average of its neighborhood. It is shown that the weight of pixel in each neighborhood decreases monotonically with the distance between it and the center pixel. Such performance is very significant because the edge is a kind of image local feature, if the smoothing operator still has significant influence on the pixel far from the center, the distortion of the image will be caused after smoothness.

Second, rotation symmetry is the feature of two-dimensional Gaussian functions. There is the same smoothness degree in a different direction. In general, the edge of an image is unknown in advance; therefore, it is not possible to determine whether more smoothing is required in one direction than in the other direction before filtering. Rotation symmetry means not to deflect in any direction during the subsequent edge detection.

Third, the dimension of Gaussian filter is decided by parameter σ. The relation is very simple between parameter σ and smoothness degree, the larger the parameter σ is the larger the smoothness degree is and the less the retained details are. It is compromised between over smoothness and less smoothness in image features in adjustment σ.

Finally, Gaussian filter is separable which can be used to implement a more complex Gaussian filter. Two-dimensional Gaussian function convolutions can be accomplished by two steps. First, the convolution is carried out between the image and one-dimensional Gaussian function, and then the convolution is carried out again between the first convolution result and the same Gaussian function of vertical direction. So, the calculation of two-dimensional Gaussian filter increases linearly with the template width rather than increasing squarely.

Gaussian filter is a widely applied smoothness filter. A series of images with different smoothness degrees can be obtained by the convolution between the image and Gaussian filter with different dimensions.

7.2.4 Stochastic Resonance

Stochastic resonance (SR) is a phenomenon that a weak signal can be amplified and optimized by the assistance of noise in bistable system, that is, this method can boost signals by noise. Its mechanism is completely different from traditional methods. In view of traditional methods mentioned above, such as Fourier transforms and Wavelet transforms, noises usually play a negative role on the detection of useful signals. These traditional methods separate the signals of interest by restraining noises rather than using of them. But they are effective only when the signals and the noise are not strongly mixed with each other in frequency spectrum. Similarly, statistical-based methods such as Wiener filter and Kalman filter, noise is also considered as the undesirable and unvalued factor. Subtracting noise is the way these methods adopted, but at the same time, the signal energy is also weakened. In contrast, SR uses noise to enhance the signal, which makes it more suitable for detecting weak electromagnetic leakage signals.

SR phenomenon can be expressed by the Langevin equation, which is governed by Eq. (7.17).

$$\frac{dx}{dt} = -\frac{\partial V(x)}{\partial x} + s(t) + \Gamma(t) \tag{7.17}$$

where $V(x)$ is the nonlinear potential function, $s(t)$ is the input signal, $\Gamma(t)$ denotes noise.

For the bistable potential, $V(x)$ can be described as:

$$V(x) = -\frac{1}{2}ax^2 + \frac{1}{4}bx^4,\ a > 0,\ b > 0 \tag{7.18}$$

Then, the governing equation can be obtained.

$$\frac{dx}{dt} = ax - bx^3 + s(t) + \Gamma(t) \tag{7.19}$$

Fig. 7.10 is a diagram of a potential function. There are two potential wells in the figure. When the energy of the signal is weak, the particles oscillate in a single-sided potential well. When noise is added, the total energy of the particles increases to achieve oscillation between the two potential wells. Under certain conditions, the oscillation frequency is the same as the signal frequency, that is, the resonance is reached.

Fig. 7.11 is a case of electromagnetic signal detection using a stochastic resonance method. After the stochastic resonance method is processed, the periodic signal in the signal is enhanced.

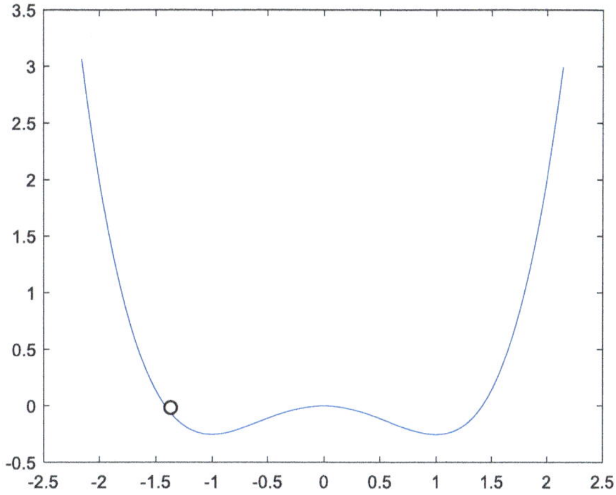

Fig. 7.10 Potential function

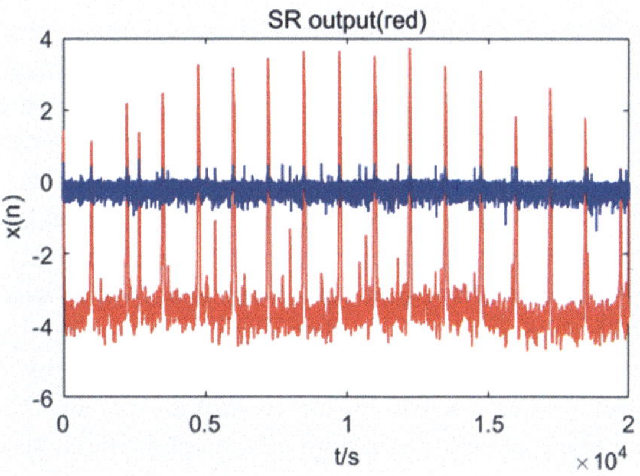

Fig. 7.11 Signal detection based on stochastic resonance

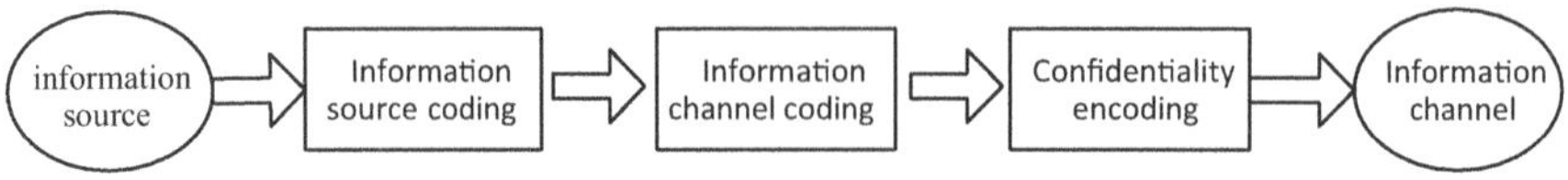

Fig. 7.12 The process of information coding

7.3 Data Process Method

7.3.1 Coding Technology

Encoding refers to a transformation of a signal for a certain purpose, the inverse of which is called decoding. Coding technology plays an important role in data processing of electromagnetic information which is significant in storage and transmission of data. Coding theory includes three branches, information source coding, information channel coding, and confidentiality coding, as shown in Fig. 7.12.

Information source coding is a transformation of signal output from the source, which includes digital-to-analog conversion and data compression. The purpose of information source coding is to improve the effectiveness of communication and it is realized by compression redundancy of information source. Normal information source coding includes matching coding, transformation coding, and so on during the information processing. Matching coding is the code length that matches the probability distribution, that is, according to the probability distribution of the object to be coded, codes of different lengths are, respectively, given. The larger the probability emerges, the shorter the given encoding length is, such as Huffman coding. Transformation encoding is to transform signal first, from one space to another space, and then the transformed signal is encoded. Transformation encoding is widely applied in speech and image encoding, such as Discrete Cosine Transformation (DCT).

Channel coding is to re-convert signal output by source encoder, in order to adapt channel conditions and improve communication reliability. Because the purpose of information source coding is to improve the information transmission reliability, the redundancy of information source is increased to realize the purpose normally which is opposite the information coding such as error detecting code and error correcting code. Error detecting code can find out the error in receiver code such as parity check code. Error correcting code can self-correct mistake to decode after receiving the error code which is an important anti-interference code.

Confidentiality encoding is to retransform the signal output from channel encoder in order to protect the signal from eavesdropping during communication which is realized by encryption data and decryption. Confidentiality encoding is recoded in order to hide the sensitivity information and prevent the information from interception and reappearance which normally adopt the methods of substitute or confusion and so on. Confidentiality encoding involves DES, AES, RSA, and so on.

7.3.2 *Image Processing Method*

In many occasions, the ultimate goal of data reconstruction and reproduction from electromagnetic information leakage is to reconstruct the original image. Therefore, using appropriate image reconstruction techniques will improve the efficiency of reconstruction, and perfect the quality of reconstructed images. Image reconstruction can use techniques and methods such as nonuniform interpolation, frequency-domain method, iterative back projection technology, and so on.

1. Nonuniform Interpolation Methods

The principle of nonuniform interpolation methods is that the image of real space is regarded as continuous function and difference image with low resolution is regarded as the sampling in different positions of continuous function, and then the nonuniform sampling is interpolated to the uniform sampling during the process of image reconstruction, and the sampling density of these sampling point is higher than that of low resolution.

The high-resolution image is reconstructed by nonuniform interpolation methods. First, information registration is accomplished inside the pixels within the images of low resolution. Then, the value is determined in location of high-resolution image point of registration information. Finally, the reconstruction is finished.

2. Frequency-Domain Method

The frequency-domain method mainly utilizes two sets of relations to restore the spectrum of original image, that is, the relationship between the continuous spectrum and the discrete spectrum, and the correspondence between the global translation parameters on space domain and the phase terms on frequency domain.

Because the spectrum aliasing is caused by the undersampling, so that the image resolution is reduced. The key of reproduction such image is to reproduce the aliased spectrum information which is the high-frequency information that also expresses details of the image.

3. Iterative Reverse Projection Techniques

The principle of iterative reverse projection techniques is that the relevant high resolution will be obtained with the error convergence when the error is projected to the high-resolution image. The low-resolution image is simulation output by the reproduced high-resolution image under the observation model of low resolution if the reproduced high-resolution image is closer to the original high-resolution image.

In addition to above methods, the image can be reproduced from various angles such as projection convex set method, probabilistic reconstruction methods, and so on. Each of the above methods has their advantages and limitations, so kinds of methods are often used together according to the characters of target image in actual work.

The processing of electromagnetic information adopts relevant process methods according to different stages of generation, radiation, and reception of electromagnetic information. Moreover, different processing technologies are applied according to different electromagnetic information sources and channels, such as encoding, error control, cryptographic, and so on.

The processing of electromagnetic information is a complex system engineering which needs to integrate many theories of various subjects. The processing methods are developed continuously and involved in more and more fields, playing significant roles on the research of electromagnetic information leakage and protection with the development of processing methods of electromagnetic information.

7.4 Summarize of Methods Based on Layered Level Protection Strategy

Usually, the original signal usually needs to undergo some processing, such as analog-to-digital conversion, digital-to-analog conversion, time-domain–frequency-domain conversion, denoising, etc., in order to meet certain preset standards, such as satisfying the need of receiving, transmitting, and interpreting. These processes are realized by various signal process methods. The processing methods used are different according to the target.

The processed signal is operated in the data treatment by information facilities. The different data treatment methods are employed according to various information facilities and data treatment targets. The corresponding treatment methods are adopted according to different treatment objects of data reconstruction and information retrieval, as shown in Table 7.2.

Researching techniques for interception and reproduction help to improve the protection technology and increase the protection effect, which is an effective method to promote the development of protection technology.

The signal treatment and data treatment methods are applied differently according to different protection emphases based on the system protection strategy of anti-leakage, anti-intercept, and anti-retrieval during the process of electromagnetic information leakage, as shown in Table 7.3.

In order to achieve the purpose of preventing electromagnetic information leakage, it is necessary to comprehensively apply various techniques for preventing leakage of electromagnetic information from multiple dimensions. Therefore, the technologies of anti-leakage, anti-interception, and anti-recurrence are not independent of each other from a strategic point of view, and they have a complementary and synergistic relationship. One method can reflect multiple-level protection strategies and achieve multiple-level protection effect. The same protection strategy can be achieved with multiple protection methods.

Table 7.2 Intercepted reproduction methods

No	Application	Methods
1	Video image reproduction technology	By filtering, data encoding and image processing methods, getting pixel clock, horizontal synchronization, vertical synchronization, and data enable signals, then reproducing the original image
2	Keyboard reproduction technology	On the basis of the short-time Fourier transform, the keyboard data is reconstructed by signal processing methods such as edge transform, matrix scan, and modulation, and using the keyboard data encoding processing method
3	Laser printer information reconstruction technology	Through filtering and other signal processing methods, and using data encoding and image processing methods, the three synchronization signals of printing start, end, and carriage return are obtained, thereby reproducing the original image
4	Key reproduction in smart card technology	Through the signal processing method, the signal characteristic curve is obtained, and the data confidentiality coding process is analyzed so as to reproduce the key

Differential protection strategies need to be implemented through signal treatment methods and data treatment methods. These methods also have a relationship that is interrelated and mutually influential. In order to obtain the same protection target, the different protection methods can be adopted, which are interaction results of many signal processing methods and data treatment methods of protection.

Therefore, various factors should be overall considered according to detail targets and condition when the protection methods are designed according to the protection strategy. The optimization protection scheme can be realized by compromised methods to choose the detail signal treatment methods and data processing methods.

Table 7.3 Methods for prevention of electromagnetic information leakage

No	Strategy	Technology	Applied processing method
1	Leak-proof	Fourier–Gaussian TEMPEST Font Technology	Through the Fourier transform and Gaussian filtering signal processing method, the space domain is converted into frequency domain and high-frequency components are removed. And, using data processing methods, electromagnetic radiation is reduced to achieve the purpose of preventing leakage
2	Leak-proof Anti-reproduction	Pixel clock randomization	Through the signal processing and image data processing methods, the pixel clock is randomized. It is difficult to obtain signals such as pixel clock, horizontal synchronization, vertical synchronization, and data enable, so as to achieve the purpose of preventing interception and reproduction
3	Anti-interception Anti-reproduction	RGB configuration technology	Through signal processing, data encoding and image processing methods, the amount of electromagnetic radiation is minimized or constant, and the purpose of preventing interception and reproduction is achieved
4	Anti-reproduction	Image plus noise	Through signal processing, data encoding, and image processing methods, random noise is added to the original image to reduce the signal-to-noise ratio of the leaked signal and achieve the purpose of preventing reproduction
5	Leak-proof	Custom kernel design method	Through signal processing, on the basis of the specialization of the data channel, the data encoding method is used to hide the sensitive data and prevent electromagnetic leakage
6	Leak-proof	TMDS coding	Through the TMDS data encoding method, electromagnetic radiation is reduced to prevent leakage

Chapter 8
Electromagnetic Information Leakage Testing

The security level of information equipment can be assessed by testing. The correct test methods are selected according to different Device Under Testing (DUT) in the appropriate environment in order to ensure the accuracy.

8.1 A Sketch of Electromagnetic Information Leakage Test

8.1.1 Purpose of Test

Through certain methods with the aid of special equipment, the Electromagnetic Information Leakage Testing (EILT) is to obtain relative data of electromagnetic information of DUT with which to access the security of the equipment with the testing results.

Though testing process and interception process are similar in some ways, there are differences between them. The purpose of interception is to obtain all information leaked from the electromagnetic radiation of the information equipment with the purpose of data reconstruction and reproduction, while the purpose of testing is to examine whether there exists leakage risk of electromagnetic information, with the purpose of evaluating security levels of information equipment to decide whether the information equipment can be used in given security requirements [67–70].

UILT can not only evaluate the security levels of DUT, but also helps to find ways of improving the design of DUT and ways of upgrading and reconstruction of the testing equipment itself.

T. Liu and Y. Li, *Electromagnetic Information Leakage and Countermeasure Technique*,
https://doi.org/10.1007/978-981-10-4352-9_8

8.1.2 Testing Methods

The features of electromagnetic information leakage signal are analyzed by frequency domain and time domain. Testing by the way of time domain can obtain the waveform of the tested signal, with the advantage of comparing and analyzing periodic signals of the waveform and thus master the parameter of its characteristic quantity. In comparison, the advantage of testing by way of frequency domain is the timely observation of frequency spectrum of the DUT radiation so that parameters of signals such as frequency range, radiation intensity, etc., can be obtained.

Generally, various assembly units of information equipment can be tested and analyzed through three measured methods, namely, shielding isolation method, near-field detecting method, and injecting probe method.

1. Shielding Isolation Method

Shielding isolation method refers to the shielding and isolation of the component units (excluding the tested units) by the application of shielding reinforcement methods or shielding materials, such as shielding screen and shielding clothing, so as to reduce as far as possible the influence of electromagnetic radiation of other units on the tested ones.

2. Near-Field Detection Method

The near-field detection method refers to test and analysis of the radiation characteristics by putting the near-field probe close to the tested units.

3. Injecting Probe Method

Injecting probe method is a kind of card electric current transducer for testing the asymmetric and symmetric interference current in the circuits (power lines, video lines). This test method can be finished without disturbing the normal work or arrangement of DUT, without the electronic contact with the source line or changing the original circuit.

All work items during the test should be in accordance with the stipulation of TEMPEST standard in order to ensure the reliability of the test results, such as follows:

(1) Selection of antennas;
(2) Lapping methods of antenna counterpoise;
(3) Orientation of antennas;
(4) Inspection methods of antenna test system;
(5) Range of test frequency, bandwidth, and measurement time;
(6) Test environment; and
(7) Test equipment.

So, appropriate methods should be selected for the test of electromagnetic information leakage according to, in principle, the features of the DUT itself, degree of precision needed, the test equipment and environmental conditions, and so on. Accurate test results cannot be achieved without proper test methods.

With the extensive application of large-scale integration circuits and microcomputers, modern test facilities are larger in measuring range, higher in accuracy, faster in speed, and easier to be automated such as the automation of range switching, calibration, diagnosing troubles and restoration, as well as the automatic recording, data calculating, and analyzing and processing of the test results.

Now, the range of frequency measurement for the test facilities can be as low as 10–4 Hz to as high as 10–12 Hz and error is decreased to 10–13 Hz. But in different ranges of frequency, even through testing the same electric quantity, methods and test facilities applied should also be different because apart from facility error, the test error may result from imprecise theoretical basis or theoretical errors caused by approximate calculation and from methodical error caused by unsuitable methods. Moreover, the test results may also be influenced by factors such as temperature, humidity, vibration, supply voltage, electromagnetic fields, etc., of the testing environment.

8.2 Testing Environment of Electromagnetic Leakage

8.2.1 Environmental Requirement

The testing environment is required to test the electromagnetic information leakage of the information facilities. The following testing environment can be selected: open ground outdoors, shielding room, microwave anechoic chamber (nonreflecting chamber), TEM cell (symmetric or dissymmetric), and GTEM cell (dissymmetric). At present, the last three types are commonly adopted.

Generally, the environmental noise is required to be 10 dB lower than the minimum electrical level of radiated interference of EUT. At the location of EUT, when it is powered off, electromagnetic noise level of transmission radiation and environmental signal level should at least be 6 dB lower than tolerable limits. The testing environment is regarded to meet the requirements of standard limits if the sum of environmental electromagnetic field level and electromagnetic radiation level of EUT do not exceed the tolerable limits.

Additional signals will be generated from the reflection of the earth. The theoretical value of additional signals is 6 dB in CISPR elliptical area (the elliptical flat area without reflectors). Other test methods can be applied to decide whether EUT meets the standard requirements if environmental levels of the environmental field or the supply power exceed the limits of standard requirement in some frequency points of the testing range.

Some flexible testing methods are as follows:

(1) The test can be performed in a shorter distance than the specified test distance, and then convert the test results into the data of specified distance.
(2) The test of the critical frequency band can be performed at the break time of the broadcasting station and when disturbance of the industry is lower.

(3) The open field can be simulated with a semi-anechoic chamber, that is, there is only direct wave and ground-reflected wave when the electromagnetic wave is diffused.
(4) The test can be performed in a shielded room or anechoic chamber.

Note: The test results taken in the shielded room cannot be regarded as the criterion to judge whether or not the system is qualified. However, the relative emission level before and after it is influenced by ambient environment that can be observed through the results tested in the shielding room. On the basis of the observation and test taken on experimental fields, emission field strength of EUT interfered by ambient environment can be relatively accurate measured.

If wider tolerant temperature is not stipulated in the standard of EUT, normally the ambient temperature of the test field is kept between 10° and 40°. If EUT and test apparatus have been located on the test field for a long time and reached the heat balance, and if the ambient temperature slightly exceeds the range of 10°–40°, the test result is acceptable if it does not cause any controversy. But error caused by the temperature influence of EUT and apparatus must be given in order for data to be amended. Normally, any desk on which to put the test apparatus should be nonconductive.

8.2.2 Indoor Test

The indoor test is normally taken in shielding room, microwave chamber, TEM cell, and GTEM cell.

The shielding room is covered with a layer of copper sheet on the inside of the building wall to ensure that the outside microwave can't penetrate to the inside of the shielding room, and vice versa, as shown in Fig. 8.1.

Microwave chamber, also called anechoic chamber or wave-absorbing chamber, is constructed by covering the ceiling, the walls, and the ground with wedge-shaped wave-absorbing material which usually differs in length. When electromagnetic wave comes in from any direction, the vast majority of it is absorbed and little is reflected or transmitted.

Fig. 8.1 Shielding room

The applied absorbing material is featured by little surface reflection and large internal loss so that the electromagnetic wave can be sufficiently attenuated by the inside materials. Common absorbing materials include plate-mutilated medium-absorbing material and the pyramid-type absorbing material, as shown in Fig. 8.2.

With the characteristics of soft texture, lightweight, and high flexibility, the plate multilayer absorbing material is usually applied as auxiliary materials in the chamber. For instance, it can be used as temporary shielding flapper to prevent clutter emission in the shielding room or to protect the staff from the hazard of microwave.

Pyramid-type absorbing material is a kind of absorbing wave material with the characteristics of high absorbing and wide frequency range and large range of incidence angle. There isn't obvious change in the absorbing quality when the incidence angle of the electromagnetic wave changes between 0° and 70° and when the incidence of different polarized states occur.

Shielding efficiency differs with the variation of the shape of the microwave shielding room and the type of sealing. The shielding efficiency of the rectangular- and the pyramid-type shielding room is better than that of other shielding rooms. The construction diagram of some types of shielding room is shown in Fig. 8.3.

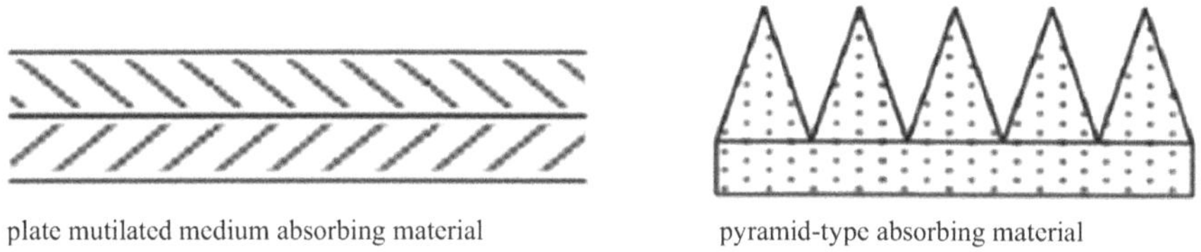

Fig. 8.2 Absorbing material

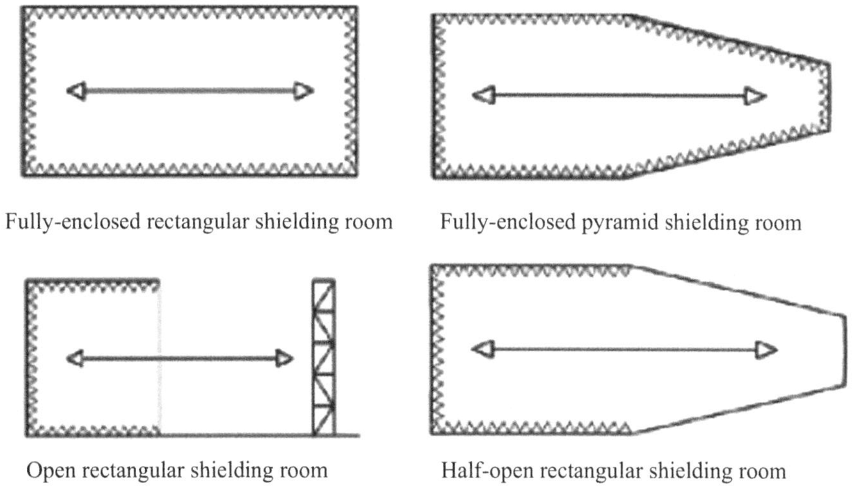

Fig. 8.3 Types of shielding room

In addition to a chamber, the microwave shielding room is installed with a supporting test and control cell for the installation of the test equipment and for testing. The test and control cell is installed on both sides of the shielding room, which is separated with the metal shielding layer from the shielding room, and the absorbing materials are not necessary for the test and control cell. The door of the shielding room should not be open to main reflection zone and the absorbing material applied on the doors should be as high as that of the surroundings. The connection between the door and the wall is sealed. Cutoff waveguide window should be applied for the ventilation opening of the microwave shielding room. Explosion-proof incandescent lamps are used for the illumination to prevent noise of electromagnetic interference.

The shielding effectiveness of microwave shielding room is 80–100 dB when the range is between 10 and 18 kHz, and the site attenuation is 30 MHz–1 GHz. The ground resistance is less than 0.9 Ω.

Main indicators of electronic characteristic for the microwave shielding room are as follows:

(1) Dead zone,
(2) Reflectivity level,
(3) Cross-polarization characteristics,
(4) Multiple loss homogeneity,
(5) Field amplitude homogeneity, and
(6) Frequency range.

Figures 8.4 and 8.5 are different sizes of microwave shielding room. Figure 8.6 shows the operating state of the microwave shielding.

GTEM cell is a sealed and whole box, which can well shield interference from outside electromagnetic wave and suppress reflection from inside electromagnetic wave depending on the absorbing materials. GTEM cell has excellent matching performance because of the gradual transition of its construction. If the EUT is the radiation source, the electromagnetic wave radiation in a different frequency band released by EUT will be concentrated to spread in the gradual changing space by GTEM cell, and shown on the oscilloscope or the spectrometer provided that EUT is located in the effective work rang of GTEM cell (EUT is located 1/3 height of the box to the top and the bottom of the case). Figure 8.7 shows the sketch map of GTEM cell.

8.2.3 Outdoor Measurement

The requirement for outdoor measurement is to guarantee the authenticity and repeatability of the test results measured in the test field. In principle, the test should be operated in the open outdoor eligible field. In special circumstances, factory building or the laboratory can also be applied. On-site measurement is only for EUT needed to be operated in the workplace.

Fig. 8.4 Small-scale test shielding room

Fig. 8.5 Large-scale test shielding room

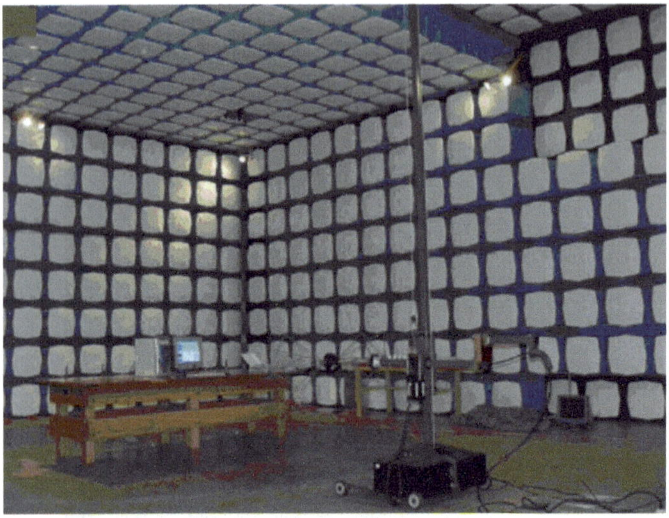

Fig. 8.6 Operating state of the microwave shielding

Fig. 8.7 The sketch map of GTEM cell

Test of the electromagnetic information leakage should be taken on open, flat, or barrier-free square. For the requirement of the open test field, there should be no reflector in the big oval field or electromagnetic pollution caused by electromagnetic reflection. The oval field is called CISPR oval. The oval border is stipulated by the distance D between EUT and the antenna and there should be no reflection from other objects within the borders.

8.3 Test Equipment of Electromagnetic Information Leakage

Test equipment of electromagnetic information leakage mainly includes receiving device and analytical instrument. Generally, the receiving device includes the receiving antenna and injection probe while the analytical instrument includes TEMPEST test receiver, digital oscilloscope, spectrum analyzer, etc. [71–73].

8.3.1 Receiver Device

The measurement of radiation leakage can be realized by the antenna, which converts the electromagnetic energy in the space into the voltage value. The antenna, according to the different functions, can be classified into the magnetic antenna and the electronic antenna. The magnetic antenna is used to receive magnetic field from the leakage during the test of EUT and magnetic field from electromagnetic environment, etc. The measurement frequency band is normally from 25 Hz to 30 MHz. The electronic antenna is used to receive electronic field from leakage during the operation of EUT and environmental electromagnetic field and so on. The measurement frequency band is normally from 10 kHz to 40 GHz. The appropriate antenna should be chosen according to the needs of the test.

There are many kinds of antennas and each antenna has its own features and receiving capability. For instance, one single antenna can't meet the requirement if the frequency band from 25 Hz to 18 GHz has to be covered for the test of information leakage. Therefore, the receiving composite unit needs to be constructed for the covering of the whole frequency band. The main parameters to evaluate antenna features are as the following: input impedance, antenna factor (AF), antenna gain (AG), and the voltage standing-wave ratio. Suitable antennas are to be chosen according to test needs and parameter requirement. The common receiving antennas mainly include low-noise active antennas, active loop antennas, passive loop antennas, high-gain horn antenna, and so on.

The injection probe is a cassette electric current sensor for measuring asymmetric interference of electric current in signal lines or power lines, and does not require electronic contact with the power wire under test nor does it require changing of the construction of electric circuit when being measured. It can be used for the measurement of conduction radiation of complicated line system and electronic circuit without disturbing the normal routine work or normal arrangement. The injection probe is also applied in measurement and analysis of the radiation signals from the equipment cables because the cable is one of the main pathways for electromagnetic information leakage.

8.3.2 Analytical Instrument

The analytical instrument includes signal analyzer, frequency measuring instrument, wave characteristic test instrument, and auxiliary instrument.

The signal analyzer is utilized to observe, analyze, and record various electrical parameter variations, such as various oscilloscopes, wave analyzer, frequency analyzer, and so on. Frequency measuring instrument is used to test frequency of electric signal, time interval, and phase difference, such as various frequency meters and phase meters. Radio wave characteristic test instrument is used to test the parameter of radio wave propagation and interference intensity and so on, such as TMEPEST test receiver, field strength meter, interference tester, and so on. The auxiliary instrument is used to coordinate with the above instruments to amplify, demodulate, isolate, and attenuate the signals so as to take maximum advantage of the above instrument, such as primary amplifier and so on.

1. Digital Oscilloscope

Currently, the main time domain analysis equipment is digital oscilloscope. With the features of the rapid and higher sampling rate, it can rapidly convert the analog signals transmitted from the antenna into digital signal, possessing the processing power of triggering waveform, memory storage, display, measurement, and waveform digital analysis.

Main technology parameters of the oscilloscope are as follows:

(1) Highest Sampling Rate

The basic operation principle of the digital oscilloscope is to sample analog signals under test and express the entire waveform with limited sampling points. The highest sampling rate means the sampling times within unit intervals, measured by the highest numbers of transmission of A/D finished per second. Sampling points/second (Sa/s) is used as its unit, and is often expressed by frequency too. The higher the sampling ratio is, the stronger the signals acquisition ability of the oscilloscope becomes. The sampling ratio is decided by A/D transition ratio. The sampling ratio of modern digital oscilloscope can be as high as 20 GSa/s.

(2) Memory Bandwidth

Memory bandwidth of digital oscilloscope is decided by the front-end hardware (input probe, etc.) and the maximum conversion rate of A/D converter. Memory bandwidth mainly reflects the maximum sampling rate and quantification accuracy (but). Maximum memory bandwidth is decided by the sampling theorem. So, when sampling rate is twice more than the frequency component of the maximum frequency of the tested signals, the sampling signal will be returned to original analog signals undistorted. Normally, there is the harmonic component in the signal, and the maximum sampling rate dividing 25 is used as the effective memory bandwidth.

(3) Resolving Ability

Resolving ability means it can differentiate minimum voltage increment and minimum time increment, or the quantitative minimum unit, including vertical-resolving ability (voltage-resolving ability) and horizontal-resolving ability (time-resolving ability). Vertical-resolving ability corresponds to the resolving ability of A/D converter, which is often expressed by the hierarchical number of screen mesh (grade/div) or percentage, or output bits of A/D converter. Presently, vertical-resolving ability of digital oscilloscope can be up to 12–14 bits. The time-resolving ability is decided by the transmission rate of A/D converter and expressed by the numbers of sampling points of each screen mesh or percentage. The accuracy and speed of A/D converter contradict with each other, so a compromise value is usually selected from them.

In addition to the above three parameters, some parameters of the oscilloscope are worth considering such as memory capacity (also called memory depth), reading speed, frequency bandwidth, rise time, input impedance, scanning speed, programmatic interface, and so on. The frequency bandwidth decides the maximum frequency $f_{\max}$ of the observable tested signal or the minimum width of a pulse signal. Only when the frequency width is three times more than $f_{\max}$ can the display without attenuation in amplitude be obtained. The rising time of the oscilloscope shouldn't be larger than 1/3 that of tested signals, so that the rising edge of pulse signals can be noted.

2. Frequency Spectrum Analyzer

Frequency spectrum analyzer can be applied in the measurement of frequency spectrum for its accurate measurement of frequency and amplitude, including distortion, stray and phase noise, and so on. Main operational features of the frequency spectrum analyzer are as follows:

(1) Frequency Width Scanning and Time Analyzing

Frequency width scanning is also called spectrum width analyzing, which displays the frequency range during one course of measurement and analysis (or one of scanning traces). The time required to finish one frequency analysis is referred to as the analysis time. The ratio of the scanning frequency width and the analysis time is referred to as the scanning frequency speed.

(2) Frequency-Resolving Ability

Frequency-resolving ability refers to the minimum line frequency interval that the spectrometer can distinguish, which indicates its capability of resolving similar frequency signals. The spectral line of the tested signals observed on the screen of the spectrometer is in fact the curve figure of a dynamic amplitude–frequency characteristic from a narrow-band filter. Therefore, the resolving ability depends on the bandwidth of the amplitude–frequency characteristic. 3 dB bandwidth of the amplitude–frequency characteristic means the resolving ability of the spectrometer. But as the curve shape of narrow-band filter of amplitude–frequency characteristic relates to the scanning frequency speed, so the resolving ability is also connected with the scanning frequency speed.

(3) Sensitivity and Dynamic Range

Sensitivity of the spectrometer refers to the ability of displaying the tiny signals when measuring the optimum resolving bandwidth. The level value of input signal when the display amplitude is full scale is called sensitivity of the spectrometer. The sensitivity depends on the noise inside the instrument. The signal spectrum line is shown above the noise spectrum especially when diminutive signals are measured. So, generally the signal level must be 10 dB more than the inside noise in order to observe clearly the signal spectral line in the noise spectrum. In addition, the sensitivity relates to the scanning frequency speed during the process of the scanning operation, so the faster the scanning speed, the lower the peak value of the dynamic amplitude frequency (the blunter of the curve), which leads to the lower value of the sensitivity and error of the amplitude value.

The dynamic range of the spectrometer shows its ability to display simultaneously the fat signal and tiny signal of the real spectrum. Upper limit of the dynamic range is restricted by the nonlinear distortion, generally more than 60 dB, or sometimes up to 90 dB. The display of the amplitude value for the spectrometer is classified into two types, namely, the linear type and the logarithm type. The logarithm type is usually adopted in order to obtain larger dynamic range from the range of limited screen altitude.

3. TEMPEST Test Receiver

Tempest test receiver is the frequency-domain measurement instrument with higher performance. Its operation principle is the following: first, the received signal is amplified and then put into the medium frequency amplifier after several grades of mixing. After that, the amplified medium frequency signal enters detector. Finally, it is output and displayed after being amplified by low frequency. As the front-end instrument of TEMPEST test, it depended mostly on the performance of the receiver to decide whether or not desired signals can be received. The received signal of tempest receiver is the precondition for other signals to be treated and analyzed.

Usually, the received signal by tempest receiver is slightly weaker in intensity, for the command pulse and the clock signal correlate with each other. Besides, types of signal are complex where the desired red signal and black undesired signal are always mixed together in terms of time and frequency and there are many modulation types. Furthermore, the red signal with information is always broadband signal occupying some bandwidth.

It is comparatively difficult for common receivers to extract desired information from these signals, but TEMPEST receiver can realize the function. Such receivers are mostly the combination of simulation type and digit type. The input signal is sent to the digital processor to finish the digitization after being handled by simulation. Digital signals are easily to be stored, sorted, and recognized. Due to the complexity of signals, it is required for the receiver to have wider receiving bandwidth, higher sensitivity, resolving ability, and larger dynamic range as well as the feature of handling real-time huge quantity of received signals and converting signals rapidly without losing any information (such as frequency and phase position).

TEMPEST receiver must meet the following requirements:

(1) Enough Sensitivity

Usually, the desired information is radiated from electromagnetic information leakage with wider bandwidth and weaker amplitude, and the minimum signals tested are usually several microvolt/meters in terms of order of magnitude. The field strength E depends on the receiver port voltage U and the antenna calibration coefficient CF, namely, E = U + CF. Therefore, the sensitivity of the receiver must be high enough for the minimum of the field intensity to be tested.

The equivalent noise power of the actual receiver can be expressed by the following formula:

$$N_{in} = KT_o B_N F \tag{8.1}$$

In the above formula (8.1), K is blitzing constant, $K = 1.37 \times 10^{-23}$ J/K; T_o is room temperature, $T_o = 29$ K; B_N is the equivalent noise bandwidth (Hz) of the receiver, which is related to intermediate frequency bandwidth and video bandwidth; F stands for noise coefficient of the receiver.

Accordingly, the sensitivity of the receiver is related to the receiver bandwidth and noise coefficient F. The noise coefficient can be decreased by low-noise amplifier located on the front end of the receiver. The noise coefficient of TEMPEST receiver can be less than 8 dB. Besides increasing the sensitivity of the receiver, receiving antennas with low noise and high gain should be applied.

(2) Intermediate Bandwidth with Wide Enough and Good Rectangle Coefficient

In order to receive complete signals, wide enough receiving bandwidth is required. Presently, the widest intermediate bandwidth is up to 500 dB. The rectangle coefficient of the filter should be better for the purpose of decreasing noise and improving selection. The quality of the filter directly affects the sensitivity of the receiver. The analog filter is used to handle continuous signals while the digital filter is used to deal with discrete signals. The features of the digital filter are closer to ideal filters than those of the analog filter. Self-adapting filter can also be designed by the digital filter with the features of self-adjusting filter parameter, which can filter certain or random signals and vulnerable or time-variant signals.

(3) Various Demodulation and Detection Methods

The electromagnetic radiation of information facilities is complex, especially the command and control system of weaponry. Therefore, in order to extract different modulation types of signals, the receiver should configure various demodulation and detection methods and output ports, such as CW, AM, FM, SSB, LOG, AM/AGC, etc., so as to extract different modulating signals and fulfill the need of varied work.

(4) Higher Automation Degree

Radiation of electromagnetic information leakage is very complex and random because various signals are mixed together and desired signals are mostly weak.

It is necessary to test process lots of signals and analyze the frequency domain and time domain in order to obtain the desired signals. Voltage, electric currency waveform, and phase position of signals can be obtained from the time domain such as pulse repetition frequency (period), amplitude rising edge and falling edge, and so on. The signals in the frequency domain can be analyzed and processed by Fourier transform through frequency spectrum. It is absolutely impossible for the analysis and processing work to be finished by hand. Hence, only when closely combined with the computer can the signals be automatically tested and processed. Generally, GPIB program can help to realize automatic test and processing.

Presently, products of TEMPEST test and receiving system include traditional pattern and search pattern according to working principle. Traditional patterns of receivers are higher in sensitivity and measuring accuracy, but more complex in measuring process and slower in speed. Typical products include DS1110 from DSI Company in USA, 8900 from WJ Company and RMT from France. Search patterns of receivers are faster in measuring speed and more flexible and convenient in operation besides its function of broadband spectrum analysis. Typical products are 71910A/P and E3830 from Agilent Company in USA, FSET-7 and FSU 8 from R&S Co., in Germany.

4. Primary Amplifier

The dynamic range and sensitivity of the receiver can be enhanced by combining the primary amplifier with the measurement receiver. The radio-frequency signal can be input into the primary amplifier after it is received by antennas for the signals to go through pretreatments like attenuation, amplification, and filtration. Finally, it is output into the receiver which is then controlled by a measurement receiver through the control interface. There are many filters like high-pass, low-pass, and band-pass filters which cause little insertion loss and can suppress strong signal input without influencing sensitivity. Signal primary amplifier can be switched so as to be matched with the receiver's input according to the range of the tested signal. Performance index of primary amplifiers mainly includes frequency range, gain value, flatness degree, etc.

In addition to the above traditional facilities, there is a virtual instrument in common use. Usually, the principal functions of electric test facility consist of three parts, such as the collection and control of signals, analysis and treatment of signals, and expression and output of the results. Traditional instruments work by means of hardware functional modules and solidified software, while for virtual instruments, the last two parts, namely, analysis and treatment of signals and expression and output of the result, are finished by a computer, with some more data collection hardware added, thus a measurement instrument based on computer is constructed as illustrated in Fig. 8.8.

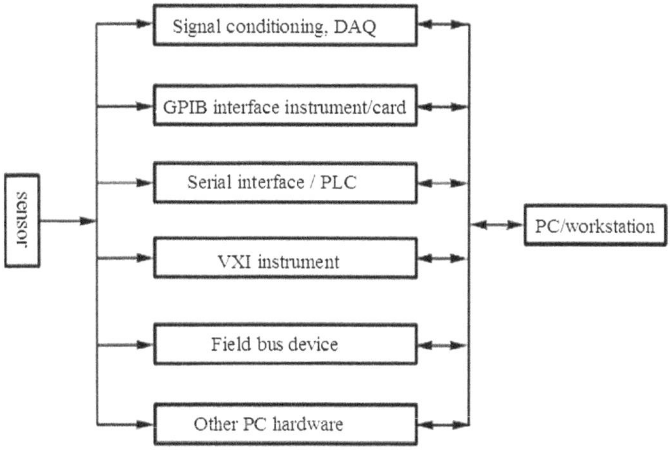

Fig. 8.8 Construction of virtual instrument

According to the different bus ways adopted by measurement function hardware, virtual instrument can be classified into the following four types:

(1) GPIB bus way of a virtual instrument.

General Purpose Instrument Bus (GPIB) was proposed by HP Co., of USA in 1965 (the predecessor of Agilent Co.,). The emergence of GPIB drives the electric measurement from the independent manual operation in a single instrument to large-scale automatic test system. GPIB technology makes it possible for a computer to operate and control a test instrument and replace the traditional manual operation pattern. Automated test system can be easily constructed by combining many instruments through GPIB port. It is most suitable for application when high accuracy is needed without the consideration of high transmission speed through computer. GPIB technology is an early development stage of a virtual instrument based on Standard EEE488.

(2) VXI bus way of a virtual instrument.

Pointed out by HP of USA and Tektronix of USA in 1987, VXI bus standard modularizes the test platform and is an expansion of VEM bus of high-speed computer in the field of virtual instrument. It inherits the feature of the GPIB bus “easy to use” and the feature of the VME bus “high handling capability” and above all, it is programmable with flexible construction and numerable channels. After 10 years’ development, VXI instrument is in wide use and more advantageous than other instruments in that the construction and application of VXI system become increasingly convenient, especially in construction of large- and medium-scale automatic measurement system as well as on occasions where higher speed and accuracy are required. However, it costs more to construct VXI bus including the cabin and the zero slot manager and the embedded-type controller.

(3) PXI bus way of virtual instrument.

PXI bus standard is pointed out by NI Co., USA in 1997. PXI is especially manufactured as a modularize instrument platform in industrious data collection and automatic application and an extension of the PCI bus in the field of the instrument. Size of PXI modules instrument is about 1/4 less than that of VXI module. Both VXI and PXI are required to configure the embedded computer in the 0 slot or the 0 slot of MXI or connect PC from outside through 1394 bus.

(4) LXI bus way of a virtual instrument.

LXI bus standard is pointed out together by Agilent Co., and VXI Co., USA in 1984. The LXI adapt to the development tendency of allocating the standard high-speed Ethernet (LAN) and the serial bus (USB) in desk computer and portable computer. Meanwhile, it is integrated with built-in measurement technology of table instrument and connective ability of PC I/O standard and the technology of the modularization and small size based on plug-in card frame systems. LXI bus standard has the features of high performance, small volume, and high-speed data throughput rate, which is a new generation of modularization platform standard based on LAN and suitable for automobile test system.

8.4 Description of Electronic Information Leakage Test System

8.4.1 Influence Factors of Test

The test result is mainly influenced by antenna system, test measurement environment, test facility, test methods, and so on.

1. Antenna System

The antenna system includes antenna selection, lapping methods of antenna earth mat, antenna orientation, and examination methods of antenna test system. Electromagnetic energy can be converted to the voltage value by antenna for the purpose of measuring radiation leakage. The antenna is divided into two kinds according to its application. One is magnetic field antenna and the other is the electric field antenna. The magnetic field antenna is used for receiving the leakage magnetic field during the operation of DUT and captivating field of space electromagnetic environment and so on. The electric field antenna is used for receiving the leakage electric field during the operation of DUT, environmental electromagnetic field, and so on. Suitable receiving antennas should be selected according to the requirement of test projects. There are many kinds of antennas for receiving electromagnetic signals with different features and receiving capacities. When a single antenna can't meet the needs, a receiving assembly unit can be constructed to cover full bands.

T property is expressed as comprehensive characters of antenna system. T represents the capability of inducting weak electromagnetic signals in the space, such as field strength sensitivity, limit test frequency, and so on. The value of T property is the weighted average of numerous performance indexes. For antenna system, the improvement of antenna sensitivity mainly depends on the development of construction, technology, and material. The performance of antennas is continuously improving with the maturity of industrial art and the development of the material science.

2. Test Environment

Test environment of electromagnetic information leakage has an obligation to meet the needs of test sensitivity and test accuracy. The alternatives of test environment include outdoor open field, the shielding room, the anechoic chamber, TEM cell, and GTEM cell. It is very difficult to find a clean open test field with the wide application of digital facility in modern society. Therefore, outdoor open field is commonly substituted by electromagnetic shielding anechoic chamber.

Combination property of test environment is expressed by property E. The property E represents the capacity of ensuring test sensitivity and accuracy and eliminating or decreasing outside interference to minimum degree during the test procedure. The value of property E is the weighted average of many performance indexes of the test environment.

3. Test Facility

There are countless types of alternative test facilities in electromagnetic test, such as the signal analyzer for observing, analyzing, and recording various electric quantity changes; the frequency measurement instrument for measuring frequency of the electric signal and time interval and phase difference; and the radio wave characteristic test instrument for measuring the parameters of the radio wave propagation, interference intensity, and so on. Other auxiliary instruments coordinate with the above various instruments for enlarging, demodulating, isolating, and attenuating signals so that the above instruments can be made best used of. Currently, regular test instruments include primary amplifier, TEMPEST test receiver, digital oscilloscope, frequency analyzer, and virtual instrument.

Combination property of test facility is expressed by property C. The property C represents the capacity of processing antenna induction signals, such as attenuation, enlargement, filtering, quantization, encoding, displaying, and preserving of the signals. The value of property C is the weighted average of various performance parameters of the test instruments.

4. Test Methods

Generally, the test methods of various components of information facilities include shielding isolation method, near-field detection method, and injection probe method or the compound modes. Different test methods will affect indexes of accuracy and precision of the test results.

Property M expresses the efficiency of the test methods, which refers to the efficiency of obtaining radiation signals intercepted from EUT. The value of the property M is the weighted average of the grade of maturity for the test facilities.

Test methods are closely related to antenna systems, test environment, and test facilities. Test methods are continuously innovated with the improvement of impact factors T, E, C.

8.4.2 Mathematical Description of Test Results

Experiments through analyzing impact factors of test results and verification through actual tests indicate that test results are affected mainly by factors like antenna system, test environment, test facility, test methods, etc. So for certain EUT, the test results can be expressed with the following models:

$$X = \varphi(T, E, C, M)|R \tag{8.2}$$

In this formula, X is the test parameters, R is radiation value of EUT, T is the property of an antenna system, E is the property of the test environment, C is the property of the test facility, and M is the property of the test method.

Formula (8.2) is expressing the measurement of the radiation quantity in the case of a given EUT, the test result X is decided by T, E, C, M, which means the accuracy of the test results is affected by the property T of antenna system, the property E of test environment, the property C of the test facility, and the property M of the test methods.

The electromagnetic radiation quantity R is decided by manufacturing technique of the equipment and operation voltage under the condition of certain operation modes. R is a fixed value in objective existence. The purpose of test is to draw the test result X near to R, as shown in the following:

$$R = MAX(X) \tag{8.3}$$

8.5 Cases of Test

TEMPEST test standard stipulates test methods, detailed test items, and relevant test value limit of electromagnetic anti-leakage for diverse information facilities. For example, there are two main documents about test requirement for TEMPEST information facility in USA, namely, NSTISSAM TEMPEST/1-92 and TEMPEST 01-02 in CNSS memorandum. Both of the documents provide detailed test methods to verify the security performance of the facilities.

Although the above TEMPEST standards are confidential, it can be understood from the unclassified documents that, for example, the US army MIL-STD-464C requires the radiation emission value of test facilities can't exceed -105 dBm/m^2 (10.8 dBμV/m) within 1 km from the location of electromagnetic radiation of its weaponry and the frequency should be 500 kHz–40 GHz. But, TEMPEST standard requirement is much higher than that.

Taking the computer system as an example, the test process of electromagnetic information leakage is introduced as follows:

(1) For the test of the computer display, in order to decrease the influence on test results from mainframe, the component units except EUT are shielded and insulated by means of shielding and strengthening or apply such shielding materials as metal gauze or shielding cloth. The test is shown in the following Fig. 8.9.

For those EUT which are not easy to be shielded or far away from other parts, near-field probe can be used by putting it close to EUT to test and analyze its radiation characteristics, such as testing the aperture of electromagnetic leakage on the surface of computer case, or testing components with large electromagnetic radiation on the surface of a circuit board. The test is shown in Figs. 8.10 and 8.11.

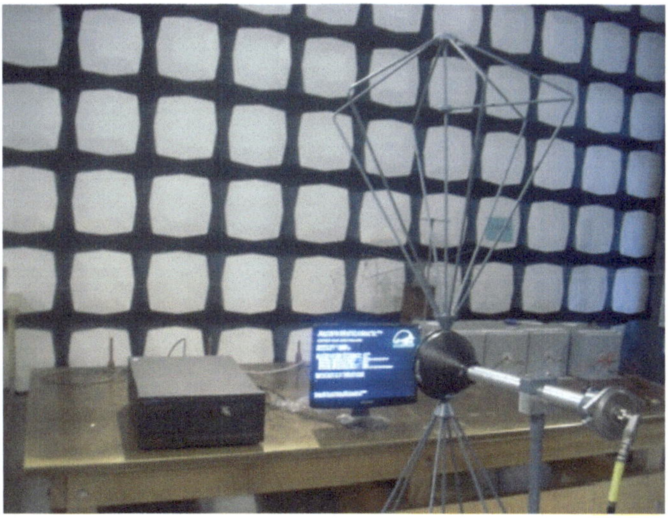

Fig. 8.9 Radiation test of the display by shielding and strengthening the mainframe

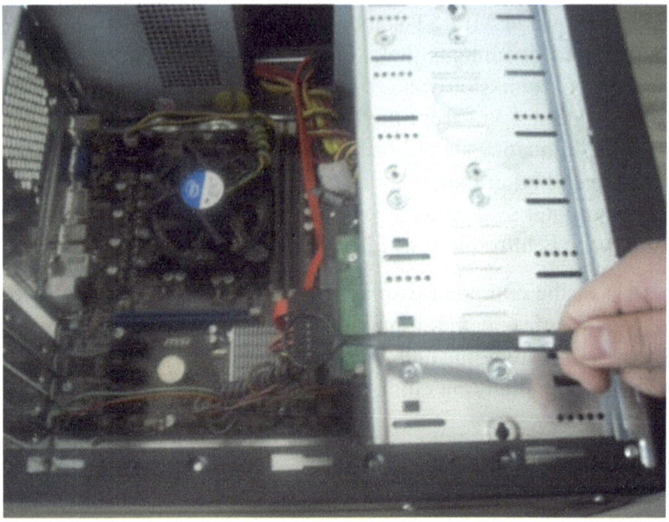

Fig. 8.10 Hardware radiation is tested by near-field probe

Fig. 8.11 Power radiation is tested by near-field probe

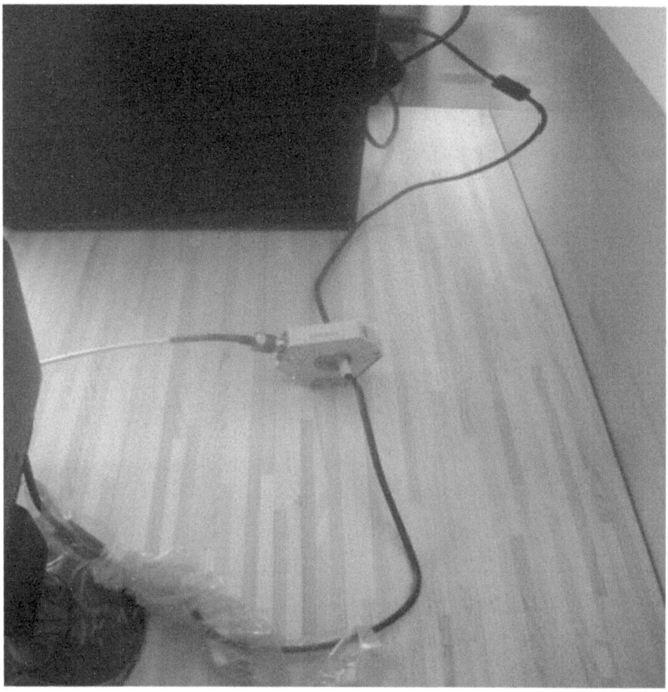

Fig. 8.12 Video line radiation is tested by injection probe

For the test of such transmission line as video line, the injection probe method can be applied to make sure that EUT can normally operate without being disturbed. The tested line should be taken as single-turn primary coil, while the secondary coil is included in the injection probe. Figure 8.12 indicates the test process.

Chapter 9
Standard Study of Electromagnetic Information Leakage and Countermeasures

TEMPEST is another research field derived from EMC technology theory. There are two goals for EMC electromagnetic compatibility standard. One is to make sure the internal circuits of electronic equipment do not interfere with one another; the other is to enable the equipment to resist external electromagnetic interference. But the goal for TEMPEST protection standard of electromagnetic information leakage is to solve the problem of electromagnetic information leakage from the information equipment, and to decrease or even prevent electromagnetic information leakage, making it difficult for the eavesdropper to intercept or reproduce the leaked electromagnetic information. So, there are connections and distinctions between TEMPEST and EMC.

9.1 TEMPEST Standard

9.1.1 About Standard

1. Property of Standard

Standard is a type of normative documents, which are authorized by a recognized institution and can be jointly and repetitively applied to obtaining optimum orders in a definite range.

The following property should be considered when setting standards, such as enforceability (cost), promotion degree (degree of acceptance, accountability, specifications, and authority), and rationality (technology). Standard needs to present the advanced technology, meanwhile, and take the cost of realization into consideration.

With the purpose of promoting optimum mutual benefit, standard is based on the comprehensive achievements of science technology and experience, which not only summarizes experience but also presents new scientific achievements as well as supporting new vital technology. The dynamic balance point should be found

T. Liu and Y. Li, *Electromagnetic Information Leakage and Countermeasure Technique*,
https://doi.org/10.1007/978-981-10-4352-9_9

Fig. 9.1 Enacting time of TEMPEST standard

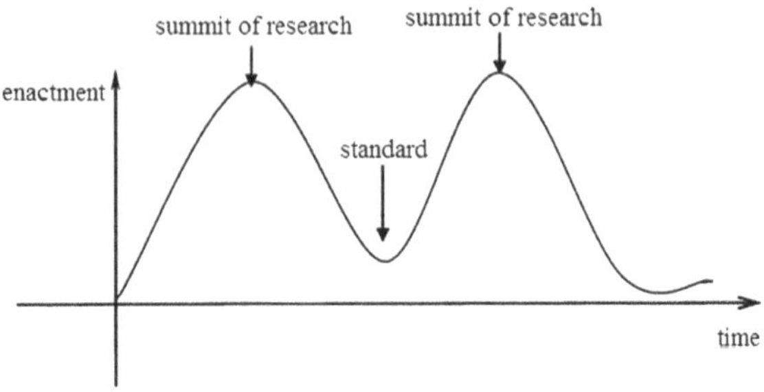

between them when setting the standard. Otherwise, the experience will be lost and the development of the latest technology will be hampered.

2. Optimum Opportunity of Setting Standard

The time for the standard to be released has to be carefully chosen, because it is vital for the standard to be successful, as shown on Fig. 9.1. The standard has released the thought between the summit of research and that of investment, which not only best reflects the research achievement but also efficiently direct the production process. If the standard is established too early, the field is still in its initial stage and the technology is immature. Inversely, if the standard is released too late, the existence of a large number of products will make it difficult to be implemented.

9.1.2 The Relationship Between TEMPEST and EMC

When electrical and electronic equipment is operated, electric current in the wire will change, which in turn generates the change of electronic field. The changed electronic field motivates the change of magnetic field, and the changed magnetic field will induce the electromotive force. Hence, electronic field and magnetic field will stimulate each other and propagate in the space. This is an inevitable natural phenomenon in actual existence. But such a phenomenon will generate two consequences: the electromagnetic interference and electromagnetic information leakage. EMC standard and TEMPEST standard is the applicable technology standard for solving the problem of electromagnetic interference and electromagnetic information leakage, respectively.

Although EMC and TEMPEST focus on solving different problems, yet there are connections and distinctions between them due to the same origin. For example, some detailed contents in the standards are overlapped, especially for military EMC standard, which is mostly related with TEMPEST standard. If TEMPEST standard is researched, EMC standard needs to be understood before studying TEMPEST.

1. EMC standard

Many organizations in the world initiate to formulate EMC Electromagnetic Compatibility standards. Some of the most authoritative and influential international standard organizations are as follows:

(1) IEC (International Electrotechnical Commission);
(2) CCIR (International Radio Consultative Committee);
(3) CISPR (International Special Committee on Radio Interference);
(4) CENELEC (the Committee of European Electrotechnical Standardization).

EMC standards formulated by these organizations are equally adopted by each country or taken as the blueprint. Now, member states of WTO have tried to impose IEC and CISPR standards as basic standards, on which EMC standard of different countries or area is formulated. EMC standards mainly involve electromagnetic compatibility and testing parameters. The detail is as the follows:

"Interference emission requirement" and "susceptibility requirement" are key requirements of the EMC standard. The former requires no electromagnetic interference be generated to the outside when the equipment is operating. The latter requires the equipment not be excessively susceptible to the outside electromagnetic interference.

Testing parameters of EMC standard mainly involve emanation radiation, conduction radiation, radiation susceptibility, and conduction susceptibility. Emanation radiation and conduction radiation are the only two approaches for electromagnetic energy to introduce into the equipment or out from the equipment; the former propagates in the space in the form of electromagnetic wave, while the latter propagates along the wire in the form of current. So these relative parameters tested and evaluated by electromagnetic compatibility standard.

In general, electromagnetic compatibility standard is divided into four levels, the construction of which is shown in Fig. 9.2. The four levels of standards are basic standards, general standards, products standards, and special products standard. The details are as the followings:

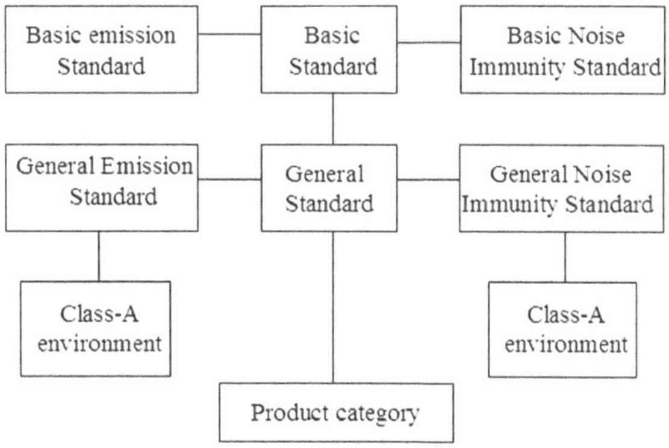

Fig. 9.2 EMC table of standard system

(1) Basic standard
The basic standards include EMC terminology, electromagnetic environment, EMC measurement equipment specifications, and EMC measurement methods, which form the basis for the development of other EMC standards.
(2) General standards
General standards provide a series of minimum electromagnetic compatibility requirements for all the products in a general environment. The testing environment and testing requirement stipulated in general standard can be utilized as the compiling rules for products standard and special products standard.
(3) Products Standard
The testing standard of EMC performance is set up according to the specific categories of products, involving electromagnetic interference emission and immunity requirement of the products.
(4) Special Products Standard
Usually, special products standard is not compiled alone as EMC standard. Instead, it is included as a special clause in the general technology specification of the products, and according to the features of the special products, the testing items can be added or certain EMC performance requirement should be modified.

Military and civilian electronic products are quite different from each other in many ways, such as in application and in performance requirement. So many countries and organizations have formulated military standards for military electronic products, namely, STANAG series standard of NATO, BS.3G.100 standard of UK military forces, VG series standard of Germany military forces, NDS standard of Japan's defense agency and EMC standard system of China, as shown in Fig. 9.3.

Generally, the military standard is stricter than the civilian standard. Taking the requirement of electromagnetic interference emission and susceptibility control in the USA, for example, the requirement of MIL-STD-461 standard is 20 times stricter than that of FCC civilian standard.

EMC standard in the USA has the maximum specifications, a complete set of equipment and is serialized. According to the EMC standard manual from the Ministry of National Defense USA (document no. ECAC-HDBK-94-088), documents about EMC standard have exceeded 500 copies. USA Military standard mainly consists of basic direction, standard, and performance specification manual. The details are shown in Appendix D.

2. TEMPEST and EMC

On account of the importance of information security and national security in strategic position, the confidentiality requirement of TEMPEST is very strict in each country, so relevant contents are strictly controlled. Therefore, it is very difficult to obtain contents of TEMPEST standard in open literature unlike EMC standard, which is openly issued. In view of the strong correlation between EMC standard and TEMPEST standard, comparison of the similarities and difference between EMC

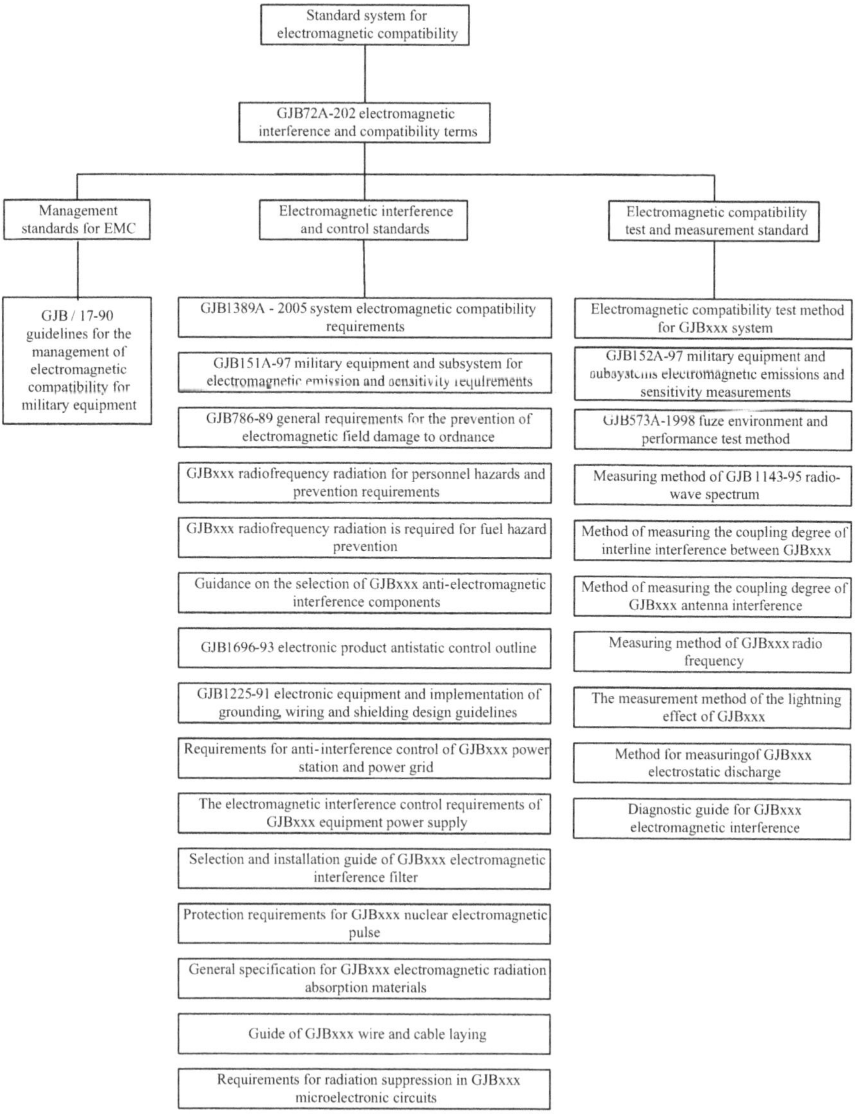

Fig. 9.3 EMC standard system in China

standard (especially military EMC standard) and TEMPEST becomes one of the effective methods to make a research on TEMPEST standard.

EMC electromagnetic standard provides standard and compulsive requirement for relevant issues, ensuring electrical and electronics equipment to operate normally with compatibility in the same electromagnetic environment without generating intolerable interference to other equipment or human bodies. Both TEMPEST standard and EMC standard aim to reduce the electromagnetic radiation, but they have different requirements and restrictions of electromagnetic radiation for different focuses. Even the military EMC standard, which mainly concentrates on the electromagnetic issues of weaponry in a special electromagnetic environment, cannot entirely solve the electromagnetic leakage issue generated by electronic equipment.

TEMPEST technology is a specialized technology for the research of electromagnetic information leakage, intercept, and reproduction. So, NSA has specially formulated TEMPEST standard on electromagnetic leakage issue, in addition to MIL-STD series standard of the USA as well as EMC departmental standards formulated by the U.S. Air Force, land force, navy force, Ministry of National Defense, Federation Aviation Administration, and Space Agency.

The purpose of formulating TEMPEST standard is to protect radiation and information leakage from being intercepted by hostile forces for fear of disclosing confidential information. Therefore, in order to stipulate an appropriate standard by analysis and assessment of hazard, not only should the features of electromagnetic leakage of our equipment be understood, but also the technological level and capacity of interception and decoding of the other side should be acquainted by all means, and moreover, it is necessary to understand threats of the operating environment of the equipment and so on.

The content of TEMPEST has something in common with EMC, both of which, in the first place, are related to electromagnetic emission and suppression measurements, so their contents are, to some extent, linked to each other. For example, the contents of TEMPEST Security are described in the U.S. MIL-STD-1542B. In the second place, the basic theory and technology in both standards are the same in that TEMPEST technology theory is built on relating EMC theory. The research of EMC was started in the 1930s and over the decades, relative methods and technology have been greatly developed, thanks to the large amount of theory research and practical work done by scientists from various countries and the highly effective international communication and cooperation.

However, the distinction between EMC and TEMPEST is also apparent [74]. They differ from each other in aspects such as research aim, specification requirement, and testing technology, etc.

(1) Different aims of research.

EMC researches the suppression of electromagnetic interference emission and enhancement of anti-jam capacity with the intention of not interfering other electronic equipment and not being interfered by other electronic equipment, while TEMPEST researches leakage, intercept, and reproduction of electromagnetic information with the intention of preventing electromagnetic information leakage and avoiding being intercepted and reproduced. As for the protection of EM information leakage, TEMPEST focuses on interrupting or suppressing EM emission of valuable information (the red signal), but pays no attention to valueless information (the black signal), and sometimes even take advantage of it to disturb the enemy's eavesdropping.

(2) Different design method.

TEMPEST adopts "means of red and black separation, which takes rigorous and standardized measures to cope with circuit, transmission, and processing of red information, in order to reduce its external coupling and radiation of EM information as much as possible. EMC focuses on the design model of aiming at all radiation of EM signals, paying its primary attention to radiation intensity without the distinction of "red and black signals".

(3) Different standard and specification.

EMC standard totally restricts EM emission interference and susceptibility, while TEMPEST standard only stipulates the requirement of EM emission of the red signal which is more rigorous than that of EMC standard.

(4) Different testing requirements.

Although both standards are similar to each other in testing, TEMPEST standard adopts stricter requirement in equipment inspection, testing field and environment, testing content, and methods, etc. Testing content involves emission field strength, stray leakage, the characteristics of red signal waveform in the time domain and frequency domain, signal bandwidth as well as the analysis of characteristics. Meanwhile, it has a higher requirement for the anechoic chamber and the shielding room.

9.1.3 TEMPEST Standard System

Since TEMPEST standard belongs to state secrets, it is impossible to know the detailed contents and requirement of TEMPEST standard of each country. Even though the TEMPEST standard document is declassified and made public, sensitive information will be deleted. Rough content of the TEMPEST standard of each country can be known only from opened file catalogs and declassified documents.

U.S.A. is the earliest country to bring forward TEMPEST standard [6]. The corresponding TEMPEST standard system can also be found in developed countries such as Britain, German, France, and Russia, etc., China also has its own standard.

1. Name and organization of the TEMPEST standard

Listed below are the names of the current TEMPEST standards of some countries and organizations. The content of TEMPEST standard can roughly be known from Tables 9.1, 9.2, 9.3, and 9.4.

2. TEMPEST Security Zone Standard

TEMPEST certification standard is first presented in the U.S.A., but the price of certified products tends to be too high for some areas where security demands are not highly required. Therefore, a new standard, known as Zone Standard, which was brought out afterward, helps to lower the security cost to some degree because

Table 9.1 Names of current TEMPEST standards in some countries and organizations

Countries	Names of standards
NATO	SDIP series standard
The U.S.A.	CNSS/NSTISS TEMPEST series standard
Canada	CID series standard
Britain	BTR series standard
China	BMB/GGBB series standard

Table 9.2 Current TEMPEST standards of NATO

Code	Names	Release time
SDIP-27	NATO TEMPEST requirements and evaluation procedures	2005.11
SDIP-28	NATO zoning procedures	2005.11
SDIP-29	Installation of electrical equipment for the processing of classified information	2006.1
SDIP-30	Installation of electronic equipment for processing of classified data	2002
SDIP-55	TEMPEST standard	

Table 9.3 Current TEMPEST standard of the U.S.A. (adapted from CNSS 2012 catalogues)

Code	Names	Release time
7000	(CNSS) TEMPEST countermeasures for facilities (U)	2004.5
7001	(NSTISS) NONSTOP countermeasures (U)	1994.6
7002	(NSTISS) TEMPEST glossary (U)	1995.3
TEMPEST 2-91	(NSTISS) compromising emanations analysis handbook (U)	1991.12
TEMPEST 1-92	(NSTISS) compromising emanations laboratory test requirements, electromagnetics (U)	1992.12
TEMPEST 2-92	(NSTISS) procedures for TEMPEST zoning (FOUO)	1992.12
TEMPEST 1-93	(NSTISS) compromising emanations field test requirements, electromagnetics (U)	1993.8
TEMPEST 2-93	(NSTISS) rationale for compromising emanations laboratory and field test requirements, electromagnetics (U)	1993.10
TEMPEST 1-95	(NSTISS) shielded enclosures (U)	1995.1
TEMPEST 2-95	(NSTISS) RED/BLACK installation guidelines (U)	1995.12
TEMPEST 1-00	Maintenance and disposition of TEMPEST equipment	2000.12
TEMPEST 01-02	(CNSS) Non-stop evaluation standard	2002.10

Table 9.4 Current TEMPEST standard in China (adapted from China National Institute of Secrecy Science and Technology)

Code	Names
BMB1-1994	EM leakage emission limits and test methods for telephone
BMB2-1998	Examining and testing methods and safety criteria for EM leakage emission of on-site operating information device
BMB3-1999	Technical requirements and testing methods for EM shielding room tackling classified information
BMB4-2000	Technical requirements and testing methods of EM jammer
BMB5-2000	Protection requirements for EM leakage emission of on-site operating secret-related information equipment
BMB6-2001	EM leakage emission limits of cryptographic devices
BMB7-2001	Test method of EM leakage emission of cryptographic devices (general rules)
BMB7.1-2001	Test method of EM leakage emission of telephone cipher machines
BMB8-2004	Authorization requirements for testing laboratory of protection products of EM leakage emission by National Security Agency
BMB19-2006	Technical requirements and testing methods for shielding cabinet of EM leakage emission
GGBB1-1999	EM leakage emission limit of information device
GGBB2-1999	EM leakage test method for information device
BMB9.1-2007	Technical requirements and testing methods for mobile communication jammers in secrecy conference rooms
BMB9.2-2007	Installation and application Guidelines for mobile communications jammers in secrecy conference rooms

products of various security levels can be chosen on the basis of detailed security requirements. Certifications, according to the Zone Standard, are classified into three levels. The security requirement of the first level is the highest, and products of this level must go through rigorous certification, applicable only for the U.S.A. government and authorized contractors. The security requirement of the second level, though lower, must be endorsed by the government. Security degree of the third level is the lowest, generally for business purposes. The ministry of National Defense of U.S.A. is under the responsibility of product certification, called NSA TEMPEST Endorsement Program. Levels of the current standard, USA NSTISSAM/1-92 Standard, is split into level 1, level 2, and level 3, respectively.

Current parallel certification standard of NATO, NATO SDIP-27 Standard, the security level is divided into Level A, Level B, and Level C. Zone standard of Germany and Britain is similar to that of U.S.A. The equivalent classification of diverse standards is given in Table 9.5. Appendix E provides a reference for the corresponding TEMEST security equipment.

Table 9.5 Table of equipment zone levels in the U.S.A. and NATO

Versions of standard	Comprehensive level	Intermediate level	Base level
NATO SDIP-27 standard	Level A	Level B	Level C
Previous NATO laboratory standards	AMSG-720B	AMSG-788A	AMSG-784
NATO zoning standards	ZONE 0	ZONE 1	ZONE 2
USA NSTISSAM/1-92 standard	LEVEL I	LEVEL II	LEVEL III

3. Industrialization of TEMPEST standard

For a country, industrialization of TEMPEST must be promoted effectively apart from establishing TEMPEST standard. The technology progress and product upgrading of TEMPEST can be carried forward only when more organizations and enterprises join in the research field of TEMPEST.

U.S.A. invests a large amount of funds into the research of TEMPEST standard Construction, meanwhile, to keep its leading position in the world in the product development and standard certification of TEMPEST. Scientific and effective management and evaluation system of the TEMPEST has been established and industrialization of TEMPEST is realized.

In the U.S.A., NSA is in charge of promoting "TEMPEST Certification Program" and certificated products can be listed in TEMPEST product catalog (ETPL TEMPEST Product List). Now, numerous enterprises and laboratories are capable of manufacturing TEMPEST products and providing relative TEMPEST service.

In Canada, CITP (Canadian International TEMPEST Program), established in 1979, is responsible for certification management of TEMPEST, and manages and controls TEMPEST products and services to meet the demand of the government. CITP provides four kinds of service, namely TEMPEST product certification, TEMPEST service certification, TEMPEST testing service certification, and TEMPEST testing facility certification.

Products that carry a CITP Certification of Canada can also be listed into ETPL of U.S.A. Many companies conduct the business of TEMPEST industrialization and provide the relevant products and service.

In U.K., CESG (Communications-Electronics Security Group) is responsible for TEMPEST certification management and provides TEMPEST management including design guide and installation of equipment, certification standard, royal standard certification, professional signal processing and analyzing, and relative training.

9.2 Development of TEMPEST Standards in China

9.2.1 Comparison of Domestic and Overseas TEMPEST Standard

The time of developing TEMPEST research is varied in many countries, and after several waves of changes, many versions of the TEMPEST standard came into existence, among which the U.S.A. developed the most versions since it started the earliest. The development history of TEMPEST standard in different countries and the organization is shown in Tables 9.6 and 9.7.

Development status of the U.S.A. TEMPEST standard is listed in Table 9.8, which shows the U.S.A. TEMPEST technologies keep improving and perfecting all the way, and that TEMPEST technology standard and stipulations have been modified and supplemented for several times during the last few decades. Besides, in terms of the name of the standard, the connotation of TEMPEST is gradually extended. Previous names of Standards were entitled with "NACS" (abbreviation for National COMSEC). Most of these standards were abolished after the 1990s, and current

Table 9.6 Development history of different TEMPEST standards

Publishers	Issuing time	Times of revision
America	1950s	6
NATO	1982	2
China	1994	1

Table 9.7 Development status of NATO TEMPEST standard

Time	Standard issued	Remarks	Versions
1980s	AMSG 720B		First version
	AMSG 788A		
	AMSG 784		
	AMSG 799B		
	AMSG 719		
2005.11	SDIP-27	Replacing AMSG 720B AMSG 788A AMSG 784	Second version
2005.11	SDIP-28	Replacing AMSG 799B	
2006.1	SDIP-29	Replacing AMSG 719	
2002	SDIP-30		
Unknown	SDIP-55		

Table 9.8 Development status of the U.S.A. TEMPEST standard (source from CNSS 2012 catalogues issued)

Time	Standard issued	Remarks
1950s	NAG-1A	
1960s	FS222/FS222A	Replacing NAG-1A
1970s	NACSEM 5100	Replacing FS222
	NACSEM 5101	NACSEM 5100 series issued
	NACSEM 5106	
	NACSEM 5110	
	NACSEM 5204	
1980s	NACSIM 5100A	Replacing NACSEM 5100
	NACSIM 5203	
1990s	NSTISSAM TEMPEST/1-91	Replacing NACSIM 5100A
	NSTISSAM TEMPEST/1-92	Replacing NSTISSAM TEMPEST/1-91
	NSTISSAM TEMPEST/2-91	Replacing NACSEM 5106
	NSTISSAM TEMPEST/1-93	Replacing NACSEM 5110
	NSTISSAM TEMPEST/2-93	Replacing NACSEM 5101
	NSTISSAM TEMPEST/1-95	Replacing NACSEM 5204
	NSTISSAM TEMPEST/2-95	Replacing NACSIM 5203
	EMPEST 2-92	NACSEM 5100 series are almost replaced
After 2000	TEMPEST 1-00	
	TEMPEST 01-02	

names of Standard after modification are labeled with NSTISS (National Security Telecommunication and Information Systems Security), which indicates the research field of TEMPEST has expanded from former communication security to information system security. Based on documents published by NSTISSC, TEMPEST standard accounts for up to 18% of standards issued by NSTISSC. So to speak, TEMPEST plays a significant role in the U.S.A. information security.

Similarly, the customary name of NATO TEMPEST was also changed from the former standards entitled AMSG (Allied Military Security Guide) into current SDIP (Security and Evaluation Agency Doctrine and Information Publication) after its revision in 2006, which demonstrated that evaluation issue must also be taken into account rather than place sole emphasis on security.

Table 9.9 Difference of TEMPEST standard between U.S.A. and China

Content of standard	The U.S.	China
Countermeasures of TEMPEST device	I/A	N/A
Glossary terms of TEMPEST	I/A	N/A
Analysis of leakage emission	I/A	N/A
Laboratory test standard	I/A	I/A
On-site test standard	I/A	I/A
Partition standard	I/A	N/A
Shielding enclosure	I/A	N/A
Installation guide for red and black device	I/A	Only for jammers
Maintenance and handling of devices	I/A	N/A
Uninterrupted evaluation standard	I/A	N/A

By contrast, TEMPEST development in China started much later, hence apparent deficiency in system construction. Specific TEMPEST Standard is issued for certain equipment while relevant standards for equipment selection and testing evaluation have not yet published.

Table 9.9 gives detailed illustrations about the difference of TEMPEST Standard development between China and America, and it is clear that TEMPEST Standard development in China is deficient, so suitable standard for our country should be formulated according to the actual situation.

9.2.2 Suggestions on TEMPEST Standard Development of China

Standards should be continuously updated to adapt to the advance of modern technology after the standard is issued on account of the change of manifold factors such as the development of electronic warfare, complexity of EM environment of information facilities, systematization and miniaturization and integration of information facilities, etc.

TEMPEST standard mainly covers the following aspects:

(1) Confidential leakage emission analysis and testing theory.
(2) Strengthening methods and process of shielding.
(3) Emission limits of TEMPEST facilities.
(4) Testing Requirement of EM confidential leakage emission laboratory.
(5) Testing Requirement of EM confidential leakage emission site.

(6) Model selection, installation, maintenance, and treatment of TEMPSET equipment.

So far, the content of the TEMPEST system in China is still incomplete, as is obviously indicated through comparison with overseas development status of TEMPEST standard. American TEMPEST standard has been updated six times since its debut publication in 1950s, expanding its coverage from the initial communication security to the current information systems, which embodies its development status of modern technology. By contrast, TEMPEST standard in China is only updated once since its initial release in 1994. Therefore, TEMPEST standard of China has to develop a new standard.

On account of the increasing significance of information security in fields of financial, government affairs and business, construction of civil TEMPEST standard system is a new and urgent task to meet the pressing demand of a suitable civil TEMPEST standard.

In addition to the establishment of TEMPEST standard system, effective TEMPEST management system and evaluation system should also be constructed as well as the formulation of supporting TEMPEST certification program in order to ensure industrialization of TEMPEST and the unceasing development of TEMPEST technology.

In summary, although many achievements have been made through years of development, being a late starter in the construction of TEMPEST standard makes China still lag behind many developed countries in the requirement of confidentiality and espionage. Hence, TEMPEST Standard has to be constantly perfected to adapt to the development of high-new technology. Besides, protective capability of TEMPEST should be regulated and strengthened by the construction of standards, and in the meantime, rational and effective TEMPEST management rules should be established to guide the development of TEMPEST technology and industry and provide technological management support for industrialization. Furthermore, an appropriate evaluation system needs to be constructed, evaluation certification work strengthened, and requirement on the application of protective measurements for the secret-related department clearly delivered.

In future, we should consider comprehensively various factors for the update of TEMPEST standard in our country, keep an eye on development trends of overseas TEMPEST and formulate a standard conforming to the requirement of the era according to national conditions. The following suggestions are worth considering for the construction of TEMPEST standard:

(1) Information security of single equipment and overall system.
(2) Research and application of innovative material and new process.
(3) New focuses on technology progress.
(4) Diversification trend of attacking technique.
(5) The latest technology of computer security.
(6) Management organization and regulation.
(7) Product certification and evaluation system.
(8) Construction of civil TEMPEST standard system.

Appendix A
US Military EMC Standards

See Tables A.1, A.2, A.3, and A.4.

Table A.1 US defense department EMC rules

Code	Name
DoD Directive 3222.3	Department of defense electromagnetic compatibility program (EMCP)
DoD Directive 4650.1	Management and use of the radio frequency spectrum
DoD Instruction 5000.2	Defense acquisition management policies and procedures part 6, section G, electromagnetic compatibility and radio frequency management
DoD Instruction 6055.11	Protection of DoD personnel from exposure to radio frequency radiation

Table A.2 US department of defense and military EMC standards

Code	Name
DoD-STD-1463	Munitions to electromagnetic fields, requirements for evaluation of (U)
DoD-STD-2169A	High altitude electromagnetic pulse (HEMP) environment (U)
MIL-STD-220A	Method of insertion loss measurement
MIL-STD-285	Methods of attenuation measurements for enclosures, electromagnetic shielding for electronic test purposes
MIL-STD-449D	Measurement of radio frequency spectrum characteristics
MIL-STD-461D	Requirements for the control of electromagnetic interference emissions and susceptibility

(continued)

T. Liu and Y. Li, *Electromagnetic Information Leakage and Countermeasure Technique*,
https://doi.org/10.1007/978-981-10-4352-9

Table A.2 (continued)

Code	Name
MIL-STD-462D	Measurement of electromagnetic interference characteristics
MIL-STD-463A	Definitions and system of units, electromagnetic interference technology
MIL-STD-464C	Electromagnetic environmental effects requirements for systems
MIL-STD-469A	Radar engineering design requirements, electromagnetic compatibility
MIL-STD-1310F	Shipboard bonding, grounding, and other techniques for electromagnetic compatibility and safety
MIL-STD-1337B	General suppression system design requirements for portable electric hand tools (Use MIL-STD-461, MIL-STD-462)
MIL-STD-1377	Effectiveness of cable, connector, and weapon enclosure shielding and filters in precluding hazards of electromagnetic radiation to ordnance, measurement of
MIL-STD-1385B	Preclusion of ordnance hazards in electromagnetic fields, general requirements for
MIL-STD-1512	Electroexplosive subsystems, electrically initiated, design requirements and test methods
MIL-STD-1541A	Electromagnetic compatibility requirements for space systems
MIL-STD-1542B	Electromagnetic compatibility and grounding requirements for space system facilities
MIL-STD-1605	Procedures for conducting a shipboard electromagnetic interference (EMI) survey (surface ships)
MIL-STD-1757A	Lightning qualification test techniques for aerospace vehicles and hardware
MIL-STD-1795A	Lightning protection of aerospace vehicles and hardware
MIL-STD-1818A	Electromagnetic effects requirements for systems
MIL-STD-1857	Grounding, bonding, and shielding design practices

Table A.3 US EMC specifications

Code	Name
MIL-B-5087B	Bonding, electrical, and lightning, protection for aerospace systems
MIL-E-6051D	Electromagnetic compatibility requirements, systems
MIL-A-17161D	Absorber, radio frequency radiation (microwave absorbing material), general specification

Table A.4 US EMC manual

Code	Name
MIL-HDBK-235-1A	Electromagnetic (radiated) environment considerations for design and procurement of electrical and electronic equipment of subsystems and systems
MIL-HDBK-237A	Electromagnetic compatibility management guide for platforms, systems, and equipments
MIL-HDBK-241B	Design guide for electromagnetic interference (EMI) reduction in power supplies
MIL-HDBK-253	Guidance for the design and test of systems protected against the effects of electromagnetic energy
MIL-HDBK-274	Electrical grounding for aircraft safety
MIL-HDBK-293	Electronic counter-countermeasures considerations in radar systems acquisition
MIL-HDBK-294	Electronic counter-countermeasures considerations in naval communication systems
MIL-HDBK-335	Management and design guidance electromagnetic radiation hardness for air launched ordnance systems
MIL-HDBK-419A	Grounding, bonding, and shielding for electronic equipments and facilities Volume! Basic Theory, Volume 2 Applications

Appendix B
TEMPEST Equipment Classification Reference

The NATO organization divides the security level of the device and gives different partition definitions, as shown in the figure (Fig. B.1).

ZONE 0 (full level):
The highest security level, the device at this level can defend less than 1 m distance attack.

ZONE 1 (middle level):
The second highest security level, the device at this level can defend against attacks of 1–20 m distance.

ZONE 2 (basic level):
The basic security level, the device at this level can defend against a distance of 20–100 m.

Corresponding to NATO's ZONE 0, ZONE 1, ZONE 2, the United States also has corresponding requirements, namely, LEVEL-I, LEVEL-II, LEVEL-III (Tables B.1 and B.2).

T. Liu and Y. Li, *Electromagnetic Information Leakage and Countermeasure Technique*,
https://doi.org/10.1007/978-981-10-4352-9

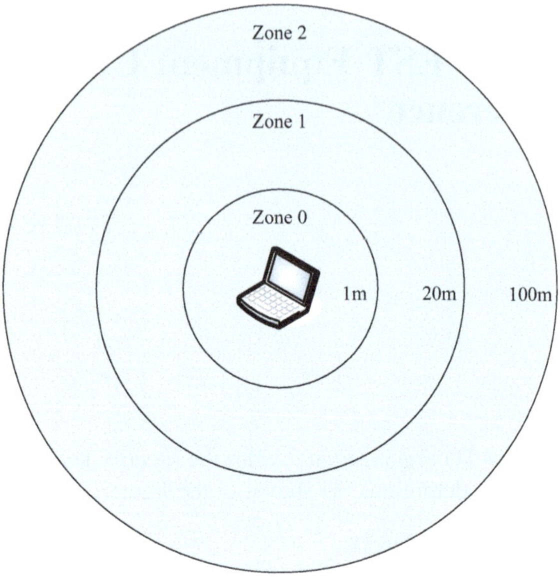

Fig. B.1 Equipment classification

Table B.1 NATO recommended product level-I

Level	Equipment	Type	Origin
I	Combination lock	X-07	US
I	Computer system	MX2-Z20	Germany
I	Computer system	MX2-Z30	
I	Computer system, pentium	YT1-PEN/90-TW3	US
I	Computer, laptop	NBTE-XXX	
I	Computer, laptop	475-T1-DL	
I	Computer, laptop	HT-430-T1-MP	
I	Computer, notebook	SSG 7770T-X	
I	Computer, personal	1486T-XX/XX	
I	Computer, personal	915-SFF1	Canada
I	Computer, personal	BTX	
I	Computer, personal	WS720	Italy
I	Computer, personal	205MTX	US
I	Computer, personal	270T-1	
I	Computer, personal	270TXA	
I	Computer, personal	274-T1-HT	
I	Computer, personal	282-T1-DL	
I	Computer, personal	SC2000BT	UK
I	Computer, personal	SC2845GBT	
I	Computer, personal	SC2865BT	
I	Computer, personal	SC2845GBT	
I	Computer, personal	646	
I	Computer, personal pentium pro	YT1-xP200-xxx	US

(continued)

Table B.1 (continued)

Level	Equipment	Type	Origin
I	Computer, rugged notebook	SN72BT-003	UK
I	Computer, rugged PC	LRT-310	Italy
I	Digital facsimile	TS-10A	US
I	Disk drive	901T-XX	
I	Display	FP18T-XXX	
I	Facsimile	SSG 890T	
I	Facsimile	SSG ILEX 795	
I	Facsimile, digital desktop	RICOH SFX2000T	
I	Fiber optic converter	QFOCV24T	UK
I	Fiber optic converter	QFOCX21T	
I	Fiber optic hub	643A	
I	Fiber optic transceiver	L2840/005T	
I	Hub	10BASE-FL Hub	US
I	Interface	SVV 8906Z	Germany
I	Interface converter	SSG 933T	UK
I	Interface, fiber optic	L2243/CT/004	
I	Interface, fiber optic	L2839/005T	
I	Interface, fiber optic	Y21881/T	
I	Interface, fiber optic	Y21882/T	
I	Interface, fiber optic	97971-Z100	Germany
I	IT system	ATS-6000	
I	Keyboard	97971-Z13x	
I	Laser printer	ATS-4200Te	
I	Modem	COMSEC LD 5010	UK
I	Modem	L2840/006T	
I	Monitor, color	T20SE-IITE	US
I	Monitor, color	TCM15	
I	Monitor, color	TCM17-X	
I	Monitor, color	DMT-107E	
I	Monitor, color	4657T	
I	Monitor, color	SL15BT-001	UK
I	Monitor, color	SL15BT-002	
I	Monitor, color	HM207A	
I	Monitor, color TFT	SL15BT-003	
I	Monitor, color TFT	SL15BT-005	
I	Monitor, LCD	DMT-101	US
I	Monitor, LCD	T15LCD-Te	
I	Monitor, LCD	4815T	
I	Monitor, LCD	4821-TI-DL	
I	Monitor, LCD	4822-T1-S	
I	Monitor, LCD	4917-T1-DL	
I	Monitor, LCD	HT-4190-T1-PH	

(continued)

Table B.1 (continued)

Level	Equipment	Type	Origin
I	Multiplexer	Y21906	UK
I	Multiplexer, fiber optic	Y21871	
I	Packet switch unit	X25T	
I	Paper tape attachment	T1560	
I	Paper tape attachment	12	
I	Printer	HC1285Z	Germany
I	Printer, black and white	HP4100-Te	US
I	Printer, dot matrix	FP 200	Norway
I	Printer, laser	LaserJetV-Te	US
I	Printer, laser	TE4050-XXXX	
I	Printer, laser	Jet DPT-103E	
I	Printer, laser	Jet DPT-104E	
I	Printer, laser	Jet DPT105e	
I	Printer, laser	915-P241-0009	Canada
I	Printer, laser	3515-T1-HP-110/3515-T1-HP-220	US
I	Printer, laser	3525-T1-HP-110/3525-T1-HP-220	
I	Printer, laser	3535-T1-HP	
I	Printer, laser	HT-34650-T1-HP	
I	Printer, laser	SC4050T	UK
I	Printer, laser	SP4100BT	
I	Printer, laser	SP4200BT	
I	Printer, laser	SP2400TF	
I	Printer, laser	SSG1100T	US
I	Printer, laser	SSG4000T	
I	Printer, laser	SSG 4049T	
I	Printer, laserjet	YT1-LJ4250-XX-C	
I	Printer, multifunction	HT-3380-T1-HP	
I	Printer, portable	SP350BT	UK
I	Rack, crypto	5020 CLD-T	US
I	Router	DNT-201e	
I	Router	SR1605BT	UK
I	Scanner	97450-T1-HP	US
I	Scanner	HT-98250-T1-HP	
I	Scanner	SC6250T	UK
I	Scanner	SC6350BT	
I	Scanner	SS7450BT	

(continued)

Table B.1 (continued)

Level	Equipment	Type	Origin
I	Scanner	SSG 6350T	US
I	Server	DST-202E	
I	Server	2700T	
I	Server	2781-T1-HT	
I	Switch	8804T	
I	Switch, ethernet	DNT-101E	
I	Switch, ethernet	DNT-301e	
I	Switch, ethernet	DNT-302e	
I	Switch, ethernet	5002T	
I	Switch, ethernet	53714-T1-AT	
I	Switch, fast ethernet	5912-T1-CS	
I	Switch, KVM	DeskSaver TC	Canada
I	Switch, KVM	8804S-T1-CY-4	US
I	Switch, thin client	TC/MLS-Te	
I	Teleprinter	DSS-9750Z	Germany
I	Teleprinter	T1501RO	UK
I	Teleprinter	T1504ESR	
I	Teleprinter	ESR615	
I	Teleprinter	610156	
I	Teleprinter keyboard	T1502KSR	
I	Terminal	LS20T	
I	Terminal	TDV 2244-Z	
I	Terminal	TDV 2270-Z	
I	Terminal, color graphics	VT370	
I	Terminal, communications	T1285Z	Germany
I	Terminal, thin client	220MC-T1-WY	US
I	Workstation	DWT-101E	
I	Workstation	DWT-104e	
I	Workstation	DWT-201	
I	Workstation	ES20Te-xx	
I	Workstation	S20Te-xx	
I	Workstation	TCQ-xxx	
I	Workstation	TWS-TE	
I	Workstation	J438T	UK
I	Workstation	XpressStation 4-Te	US
I	Workstation, portable	NB3-Te	

Table B.2 NATO recommended product level-II

Level	Equipment	Type	Origin
II	Computer, laptop	GRiDCASE 1587B	UK
II	Computer, laptop	JL/RLT/1	
II	Computer, notebook	SN72AT-002	
II	Computer, notebook	SN500AT	
II	Computer, personal	HTOUA	
II	Computer, personal	RP6000	
II	Computer, personal	EXI	
II	Computer, personal	270T-2	US
II	Computer, personal	283R-T2-HT	
II	Computer, personal	386/33T	Norway
II	Computer, personal	SC6000	UK
II	Computer, personal	SC6000AT	
II	Computer, personal	SC6845AT	
II	Computer, personal	SC6845GAT	
II	Computer, personal	SC6865AT	
II	Computer, personal	SC8000T	
II	Computer, personal	SC8800T	
II	Computer, personal	635B 386SX	
II	Computer, personal	645	
II	Computer, portable	GKM LXI Computer	
II	Computer, portable	DRS LXI UDT PUT Computer	
II	Computer, portable	GRiDCASE 1587XGA	
II	Computer, rugged laptop	LT184	
II	Computer, rugged notebook	LT1841TI	
II	Computer, rugged notebook	SN72AT	
II	Data link processor system	DLPS	
II	Display	SM15AT-007	
II	Display	SM17AT	
II	Display	DML-103E	US
II	Display	DML-105E	
II	Facsimile	TS-21T	UK
II	Facsimile	SF5780TI	
II	Facsimile	640	
II	Facsimile	640A	
II	Facsimile	650A	
II	Fiber optic dist. unit	L2839/HX-STP	
II	Fiber optic modem	FOCRS232R	
II	Fiber optic transceiver	L2840/HT-ST	
II	Hub, switched	SB8000AT	
II	Keyboard	QK102/776/R	

(continued)

Table B.2 (continued)

Level	Equipment	Type	Origin
II	Laser printer	HP 2200DN	Germany
II	Modem, fiber optic	LD-7132	UK
II	Monitor, color	L2M17-Te	US
II	Monitor, color	DML-107E	US
II	Monitor, color	4157T	US
II	Monitor, color	SL15AT-002	UK
II	Monitor, color	SM15AT-005	UK
II	Monitor, color	SM15T-001	UK
II	Monitor, color	SM15T-003	UK
II	Monitor, color	SM15T-004	UK
II	Monitor, color	SM17AT-006	UK
II	Monitor, color	HM-114B	UK
II	Monitor, color	HM-170B	UK
II	Monitor, color	SM15T-001	UK
II	Monitor, color TFT	SL15AT-005	UK
II	Monitor, color TFT	SL17AT-001	UK
II	Monitor, color TFT	SL17AT-002	UK
II	Monitor, color TFT	SL20AT-001	UK
II	Monitor, Color	HTOUB	UK
II	Monitor, high resolution	HM-174B	UK
II	Monitor, LCD	DML-101	US
II	Monitor, LCD	DML-103E	US
II	Monitor, LCD	DML-106E	US
II	Monitor, LCD	L15LCD-Te	US
II	Multiplexer/DE multiplexer fiber optic	L3201/210/ST	UK
II	Packet switch	CPX10-5T X.25	UK
II	Packet switch	CPX10-3T	UK
II	Printer	100L	UK
II	Printer	400M	UK
II	Printer	PL-119B LED	UK
II	Printer	PL-137 B6000	UK
II	Printer	636A	UK
II	Printer, color	Inkjet BLAZE 200i	UK
II	Printer, dot matrix	CP8240	UK
II	Printer, dot matrix	FP 100	Norway
II	Printer, dot matrix	280T	UK
II	Printer, dot matrix	636C	UK
II	Printer, dot matrix	636F	UK

(continued)

Table B.2 (continued)

Level	Equipment	Type	Origin
II	Printer, fiber optic	HC1285Z	Germany
II	Printer, laser	HP4100-Le	US
II	Printer, laser	6T	UK
II	Printer, laser	1152000	
II	Printer, laser	1153000	
II	Printer, laser	T612	
II	Printer, laser	SC4000T	
II	Printer, laser	SP1300AT	
II	Printer, laser	SP4100AT	
II	Printer, laser	SP4200AT	
II	Printer, laser	SP5100AT	
II	Printer, laser	SP2400TI	
II	Printer, laser	PL-154B (Parallel)	
II	Printer, laser	PL-154B (Serial)	
II	Printer, laserjet	SP4650TI-900	
II	Printer, ruggedish lightweight	V2-PT(T)	
II	Printer, thermal ink jet	SC895T	
II	Router	SC1005T	
II	Scanner	SS5400AT	
II	Scanner	YT6-3CT-FEED	US
II	Switch	YT1-SWxxx	
II	Terminal	LSX17CR	UK
II	Terminal, message	7392	
II	Terminal, window color	VXT 2000 X	
II	Terminal, window mono	VXT 2000 X	
II	Video display unit	VR319	
II	Workstation	DWL-101E	US
II	Workstation	DWL-104e	
II	Workstation	4000 90A	UK
II	Workstation	JLTS/1/P133-FO/T	
II	Workstation, desktop	LWS-TE	US
II	Workstation, TEMPEST	LWS-TE	UK

References

1. SST (2010) TEMPEST introduction [EB/OL]. http://www.sst.ws/tempest_standards.php.
2. van Eck W (1985) Electromagnetic radiation from video display units: an eavesdropping risk? Comput Secur 4:269–286
3. Smulders P (1990) The threat of information theft by reception of electromagnetic radiation from RS-232 cables. Comput Secur 9:53–58
4. Hemming LH (2000) Architectural electromagnetic shielding handbook. The Institute of Electrical and Electronics Engineers, Inc.
5. Scott BW (1997) Control and measurement of unintentional electromagnetic radiation. Wiley
6. U.S. CNSS (2012) Index of national security systems issuances. NSA, Ft. Meade, MD
7. Yinghong W, Keyen Z, Yong C, Jun C (2010) Electromagnetic field and electromagnetic compatibility. Science Press, Beijing
8. Qimin J (2011) Electromagnetics. Higher Education Press, Beijing
9. Guru BS, Hiziroglu HR (2000) Electromagnetic fields and magnetic waves. Machinery Industry Press, Beijing
10. Shizhao S, Lusheng C (2010) Information theory and coding theory. Science Press, Beijing
11. Hui Y, Zhenghong D, Weijie H (2010) Military network electromagnetic information security technology. National Defense Industry Press, Beijing
12. Jiaming P, Xueping Y (2006) Electromagnetic leakage mechanism analysis of LCD display. Electron Qual 11:65–67
13. Hayashi Y (2012) Evaluation of information leakage from cryptographic hardware via common-mode current. IEICE Trans Electron E95-C(6):1089–1097
14. Lei L, Yinghua L, Wenhan Y (2011) A new type of leaky electromagnetic wave search system. Chin J Radio Sci 26 Suppl.:123–127
15. Zou X, Yang L, Tong Y, Fang L, Shaohua X, Zhenjiang S (2000) Electromagnetic leakage in microcomputer systems. Comput Eng 26(6):59–60
16. Kinugawa M, Hayashi YI, Mizuki T, Sone H (2011) Information leakage from the unintentional emissions of an integrated RC oscillator. In: Electromagnetic compatibility of integrated circuits (EMC Compo), pp 24–28
17. Changlin Z, Zhongyi Z, Qingmei C, Mingmei G, Shaobo Z (2005) Reception and reproduction of computer electromagnetic leakage information. Electro-opt Control (2):71–73
18. Yu L, Ling L, Mingxin N, Shunliao Y, Jun L, Wenwu S, Nan W (2004) Steal mechanisms and recurring technologies of computer electromagnetic leakage information. J Wuhan Univ Technol 28(6):883–886
19. Zhiqiang Z, Taikang L, Yun J (2013) Detection and analysis of electromagnetic information leakage on computer graphics cards. Comput Secur 2:20–22

T. Liu and Y. Li, *Electromagnetic Information Leakage and Countermeasure Technique*,
https://doi.org/10.1007/978-981-10-4352-9

20. Guo L, Liu T, Ma T (2008) Electromagnetic leakage test of monitors based on EMC shielded darkroom. Comput Secur 12:43–45
21. Guo L (2009) Research on characteristic parameters of computer video signal. Taiyuan University of Science and Technology, Taiyuan
22. Zhiqiang Z (2013) Detection and analysis of electromagnetic information in computer display systems. Taiyuan University of Science and Technology, Taiyuan
23. Tosaka T, Yamanaka Y, Fukunaga K (2011) Method for determining whether or not information is contained in electromagnetic disturbance radiated from a PC display. IEEE Trans Electromagn Compat 53(2):318–324
24. Tosaka T, Ishikawa R (2007) Evaluation of information leakage from PC displays using spectrum analyzers. In: The Institute of Electronics, Information and Communication Engineers, pp 3315–3318
25. Watanabe T, Nagayoshi H, Sako H (2008) A display technique for preventing electromagnetic eavesdropping using color mixture characteristic of human eyes. In: IH, LNCS, vol 5284, pp 1–14
26. Chengdong W, Taikang L, Yun J (2012) Spectrum test and analysis of wireless keyboard radiation signal. Comput Secur (2):5–7
27. Chengdong W (2012) Research on information security of computer wireless keyboard electromagnetic radiation signal. Taiyuan University of Science and Technology, Taiyuan
28. Chen Y, Jiaming P, Xueping Y (2006) Electromagnetic information leakage analysis of computer clock signals. Electron Qual 8:67–70
29. Wang L, Yu B (2011) Analysis and measurement on the electromagnetic compromising emanations of computer keyboards, pp 640–643
30. Vuagnoux M, Pasini S (2010) An improved technique to discover compromising electromagnetic emanations. In: 2010 IEEE international symposium on electromagnetic compatibility (EMC), pp 121–126
31. Tosaka T, Taira K, Yamanaka Y, Nishikata A, Hattori M (2006) Feasibility study for reconstruction of information from near field observations of the magnetic field of a laser printer. In: 2006 17th international Zurich symposium on electromagnetic compatibility, vol 1 and 2, pp 630
32. Meynard O, Hayashi Y, Homma N, Guilley S, Danger J-L (2011) Identification of information leakage spots on a cryptographic device with an RSA processor. In: 2011 IEEE international symposium on electromagnetic compatibility (EMC), pp 773–778
33. Gandol K, Mourtel C, Olivier F (2001) Electromagnetic analysis: concrete results. In: Cryptographic hardware and embedded systems—CHES 2001, vol 2162 of Lecture notes in computer science, pp 251–261
34. Qingmei C, Changlin Z, Zhongyi Z (2000) Design and implementation of low-emission computer system. Electro-Opt Control (3):44–47
35. Yan Q, Tao L, Tao W (2004) Electromagnetic leakage and protection of the position image surveillance system. Ordnance Ind Autom (6):23–24
36. Yahong G, Baokui W (2007) Computer information electromagnetic leakage and protection. J Chang Norm Univ (3):32–36
37. Man L, Yujiang L (2003) Computer electromagnetic leakage and its protection. Aerosp Electron Warf 5:46–48
38. Haiquan L (2002) Anti-information leakage technology for computers and their external devices. Comput Eng Des 23(4):42–46
39. Ikematsu T, Hayashi Y, Mizuki T, Homma N, Aoki T, Sone H (2011) Suppression of information leakage from electronic devices based on SNR. In: 2011 IEEE international symposium electromagnetic compatibility (EMC), pp 920–924
40. Fang H (1993) Electromagnetic leakage and protection of computer information. Science Press, Beijing
41. Dali Z, Degang S, Hong D (2003) TEMPEST protection: shield for confidential information electromagnetic leakage. Inf Secur Commun Confid 2

42. Sekiguchi H (2012) Novel information leakage threat for in-put operations on touch screen monitors caused by electromagnetic noise and its countermeasure method. Prog Electromagn Res B 36:399–419
43. Ziteng F, Feng L, Yu L (2004) TEMPEST technology in computer system. J Chengdu Univ Inf Technol 19(3):418–421
44. Tanaka H (2007) Information leakage via electromagnetic emanations and evaluation of TEMPEST countermeasures. In: ICISS, LNCS, vol 4812, pp 167–179
45. Tanaka H, Takizawa O, Yamamura A (2004) Evaluation and improvement of the TEMPEST fonts. In: WISA, LNCS, vol 3325, pp 457–469
46. Kuhn MG, Anderson RJ (1998) Soft TEMPEST hidden data transmission using electromagnetic emanations. In: Information hiding 1998, LNCS, vol 1525, pp 124–142
47. Tanaka H (2008) Evaluation of information leakage via electromagnetic emanation and effectiveness of Tempest. IEICE Trans Inf Syst E91d:1439–1446
48. Tillich S, GroBschadl J (2007) Power analysis resistant AES implementation with instruction set extensions. In: Paillier P, Verbauwhede I (eds) CHES 2007, LNCS, vol 4727, pp 303–319
49. Agrawal D, Archambeault B, Rao JR, Rohatgi P (2002) The EM side-channel(s). In: CHES 2002, LNCS, vol 2523
50. Shen S-S, Chiu J-H (2008) Prevention of information leakage by photo-coupling in smart card. IEICE Trans Fundam E91-A(1):160–167
51. Jing Z (2004) Electromagnetic shielding design of shelters. Mod Electron Eng (1):78–83
52. Xiaoxiao W (2007) Study on wave-absorbing properties of absorbers for left and right hand material structures. Harbin Engineering University, Harbin
53. Haiquan L, Jian L (2001) Computer system security technology. People's Posts and Telecommunications Press, Beijing
54. Sen S (2011) Red and black signal analysis of mainframe electromagnetic leakage. Taiyuan University of Science and Technology, Taiyuan
55. Sen S, Taikang L, Yun J (2011) Test and analysis of electromagnetic leakage of computer host based on red and black signals. Comput Secur (1):44–46
56. Youxi L (2010) Research on leakage of electromagnetic information from computer hosts. Taiyuan University of Science and Technology, Taiyuan
57. Bing C (2005) Computer display TEMPEST detection technology. Taiyuan University of Science and Technology, Taiyuan
58. Li X (2007) Time domain and frequency domain analysis of electromagnetic information radiation from CRT monitors. Taiyuan University of Science and Technology, Taiyuan
59. Shuquan L (2008) Electromagnetic leakage signal analysis and processing of computer video. Taiyuan University of Science and Technology, Taiyuan
60. Kuhn MG (2004) Electromagnetic eavesdropping risks of flat-panel displays. In: PET 2004, LNCS 3424, pp 88–107
61. Kuhn MG (2005) Security limits for compromising emanations. In: CHES 2005, LNCS, vol 3659, pp 265–279
62. Xueping Y, Shuoping Y (2007) Anti-interception of electromagnetic radiation leakage from computer video information. Electron Qual 2:54–58
63. Jianqing Yao (2009) Concise digital signal processing. People's Posts and Telecommunications Press, Beijing
64. Huifen L, Xianghang J, Zhu L (2004) Research and improvement of robust performance of Gaussian filter. Chin J Sci Instrum 25(5):633–637
65. Liu S, Lu P (2012) Scanning image descreening based on Gaussian filtering. Packag Eng 13:108–111
66. Lijuan Y, Baihua Z, Xuxi Y (2004) Fast Fourier transform FFT and its application. Optoelectron Eng 31:1–3
67. Zhang H, Zhang J, Wang D, Lu Y (2012) Quantitative evaluation of information leakage arising from computer. Int J Appl Electromagn Mech 40(2):101–111

68. Yang Q, Hongting L, Weiguo S, Guangda C (2004) Research on the identification of red signal in computer video information leakage. J Mar Electron Eng 24(21):247–249
69. Dawei Z, Lei S, Haiying Z (2011) Research on TEMPEST technology of computer system. Mod Electron Technol 34(23):88–90
70. Wenbin Z, Wei F (2006) Electromagnetic shielding effectiveness test and electromagnetic leakage analysis of typical chassis. Electron Qual 11:67–69
71. Qihai S, Binhong L (2005) Electromagnetic spectrum measurement and analysis. Electron Meas Technol 3:8–9
72. Jun L, Ling L, Mingxin N, Zonggang D, Wei L, Shunliao Y (2005) Computer-based electromagnetic leakage data processing platform based on TEMPEST. J Wuhan Univ Technol 29(1):83–86
73. Changlin Z, Deting H, Qingmei C, Xunyi Z (2005) Testing and analysis of electromagnetic leakage spectrum of digital information technology equipment. Comput Eng 31(17):189–190
74. Jiemin Z, Yongmei L (2013) The study of the standards architecture and the standards attributes based on EMC standards and TEMPEST standards in computer system. In: The 8th international conference on computer science and education (ICCSE 2013), in Colombo, Sri Lanka, pp 224–227
75. Liu J, Mao J, Huang B, Liu P (2018) Phys Lett A 382:3071–3078

The manufacturer's authorised representative in the EU is Springer
Nature Customer Service Centre GmbH, Europaplatz 3, 69115 Heidelberg,
Germany. If you have any concerns regarding our products, please
contact ProductSafety@springernature.com

Printed and bound by CPI Group (UK) Ltd, Croydon, CR0 4YY
30/09/2026
02225440-0005